Enhanced Abutment Scour Studies for Compound Channels

ENHANCED ABUTMENT SCOUR STUDIES FOR COMPOUND CHANNELS

Report No. FHWA-RD-99-156

FOREWORD

This report describes a laboratory study of abutment scour for compound channels where the experiments simulated floodplains with defined channel and overbank flow areas at different elevations. A new abutment scour prediction equation was derived as part of this study and is presented in an appendix of the fourth edition of the Federal Highway Administration Hydraulic Engineering Circular No. 18 (HEC-18). This report will be of interest to bridge engineers and hydraulic engineers involved in bridge scour evaluations and to researchers involved in developing improved bridge scour evaluation procedures. It is being published as a Web document only.

T. Paul Teng, P.E.
Director, Office of Infrastructure
Research and Development

NOTICE

1. Report No. FHWA-RD-99-156	2. Government Accession No.	3. Recipient's Catalog No.
4. Title and Subtitle Enhanced Abutment Scour Studies for Compound Channels		5. Report Date August 2004
		6. Performing Organization Code
7. Author(s) Terry W. Sturm		8. Performing Organization Report No.
9. Performance Organization Name and Address Georgia Institute of Technology School of Civil and Environmental Engineering Atlanta, GA 30332		10. Work Unit No. (TRAIS)
		11. Contract or Grant No. DTFH61-94-C-00198
12. Sponsoring Agency and Address Office of Infrastructure Research and Development Federal Highway Administration 6300 Georgetown Pike McLean, VA 22101-2296		13. Type of Report and Period Covered Final Report 1994-1999
		14. Sponsoring Agency Code
15. Supplementary Notes Contracting Officer's Technical Representative (COTR): J. Sterling Jones, HRDI-07		

16. Abstract

Experimental results and analyses are given in this report on bridge abutment scour in compound channels. Experiments were conducted in a laboratory flume with a cross section consisting of a wide floodplain adjacent to a main channel. The embankment length, discharge, sediment size, and abutment shape were varied, and the resulting equilibrium scour depths were measured. Water-surface profiles, velocities, and scour-hole contours were also measured. In the report, a methodology is developed for estimating abutment scour that takes into account the redistribution of discharge in the bridge contraction, abutment shape, sediment size, and tailwater depth. The independent variables in the proposed scour formula are evaluated at the approach-channel cross section and can be obtained from a one-dimensional water-surface profile computer program such as the Water-Surface Profile Program (WSPRO). The proposed scour evaluation procedure is outlined and illustrated, including consideration of the time required to reach equilibrium scour. The proposed methodology is applied to two cases of measured scour in the field.

17. Key Words Bridge scour, abutment scour, compound channels.	18. Distribution Statement No restrictions. This document is available to the public through the National Technical Information Service, Springfield, VA 22161.		
19. Security Classif. (of this report) Unclassified	20. Security Classif. (of this page) Unclassified	21. No. of Pages 144	22. Price

SI* (MODERN METRIC) CONVERSION FACTORS

APPROXIMATE CONVERSIONS TO SI UNITS

Symbol	When You Know	Multiply By	To Find	Symbol
LENGTH				
in	inches	25.4	millimeters	mm
ft	feet	0.305	meters	m
yd	yards	0.914	meters	m
mi	miles	1.61	kilometers	km
AREA				
in^2	square inches	645.2	square millimeters	mm^2
ft^2	square feet	0.093	square meters	m^2
yd^2	square yard	0.836	square meters	m^2
ac	acres	0.405	hectares	ha
mi^2	square miles	2.59	square kilometers	km^2
VOLUME				
fl oz	fluid ounces	29.57	milliliters	mL
gal	gallons	3.785	liters	L
ft^3	cubic feet	0.028	cubic meters	m^3
yd^3	cubic yards	0.765	cubic meters	m^3
	NOTE: volumes greater than 1000 L shall be shown in m^3			
MASS				
oz	ounces	28.35	grams	g
lb	pounds	0.454	kilograms	kg
T	short tons (2000 lb)	0.907	megagrams (or "metric ton")	Mg (or "t")
TEMPERATURE (exact degrees)				
°F	Fahrenheit	5 (F-32)/9 or (F-32)/1.8	Celsius	°C
ILLUMINATION				
fc	foot-candles	10.76	lux	lx
fl	foot-Lamberts	3.426	candela/m^2	cd/m^2
FORCE and PRESSURE or STRESS				
lbf	poundforce	4.45	newtons	N
lbf/in^2	poundforce per square inch	6.89	kilopascals	kPa

APPROXIMATE CONVERSIONS FROM SI UNITS

Symbol	When You Know	Multiply By	To Find	Symbol
LENGTH				
mm	millimeters	0.039	inches	in
m	meters	3.28	feet	ft
m	meters	1.09	yards	yd
km	kilometers	0.621	miles	mi
AREA				
mm^2	square millimeters	0.0016	square inches	in^2
m^2	square meters	10.764	square feet	ft^2
m^2	square meters	1.195	square yards	yd^2
ha	hectares	2.47	acres	ac
km^2	square kilometers	0.386	square miles	mi^2
VOLUME				
mL	milliliters	0.034	fluid ounces	fl oz
L	liters	0.264	gallons	gal
m^3	cubic meters	35.314	cubic feet	ft^3
m^3	cubic meters	1.307	cubic yards	yd^3
MASS				
g	grams	0.035	ounces	oz
kg	kilograms	2.202	pounds	lb
Mg (or "t")	megagrams (or "metric ton")	1.103	short tons (2000 lb)	T
TEMPERATURE (exact degrees)				
°C	Celsius	1.8C+32	Fahrenheit	°F
ILLUMINATION				
lx	lux	0.0929	foot-candles	fc
cd/m^2	candela/m^2	0.2919	foot-Lamberts	fl
FORCE and PRESSURE or STRESS				
N	newtons	0.225	poundforce	lbf
kPa	kilopascals	0.145	poundforce per square inch	lbf/in^2

*SI is the symbol for the International System of Units. Appropriate rounding should be made to comply with Section 4 of ASTM E380.
(Revised March 2003)

TABLE OF CONTENTS

LIST OF FIGURES

v

LIST OF TABLES

LIST OF SYMBOLS

a = blocked flow area by abutment

A = unobstructed flow area in approach channel

A_{fl} = blocked flow area in approach-channel floodplain

b = bridge pier width

B_f = width of floodplain in compound channel

B_m = width of main channel in compound channel

B_{m1} = width of approach main channel

B_{m2} = width of main channel in contracted section

C_n = units conversion coefficient used in equation 23

C_0 = best-fit constant in proposed scour formula

C_r = best-fit coefficient in proposed scour formula

C_{LB} = live-bed scour coefficient

d_{50} = median diameter of sediment

d_s = local scour depth

d_{sc} = theoretical long contraction scour depth

d_{st} = unsteady scour depth at any time t

F_0 = approach value of densimetric grain Froude number in Lim scour formula

F_1 = approach-flow Froude number

F_{ab} = Froude number adjacent to abutment face

$k\text{-}\varepsilon$ = k-epsilon refers to a numerical modeling technique that uses the ratio of turbulent kinetic energy squared to turbulent energy dissipation rate to define the eddy viscosity term for computations

k_f = spiral-flow adjustment factor in Maryland scour formula

k_s = equivalent sand-grain roughness

k_v = velocity adjustment factor in Maryland scour formula

K_1 = geometric shape factor for abutment in Froehlich scour formula

K_2 = embankment skewness factor in Froehlich scour formula

K_s^* = abutment shape factor in Melville scour formula

K_G = channel geometric factor in Melville scour formula

K_I = flow-intensity factor in Melville scour formula

K_{ST} = spill-through abutment shape factor for scour formula in present study

K_θ^* = abutment alignment factor in Melville scour formula

L_a = abutment/embankment length

M = discharge distribution factor in approach section = unblocked Q over total Q
  = discharge contraction ratio for a bankline abutment

n = Manning's n

n_{fp} = Manning's n in floodplain of compound channel

n_{mc} = Manning's n in main channel of compound channel

n_K = Manning's n calculated from Keulegan's equation

N_s = sediment number (equivalent to densimetric grain Froude number)

N_{sc} = critical value of sediment number

q_1 = flow rate per unit width in approach section in Maryland scour formula

q_2 = flow rate per unit width in contracted section in Maryland scour formula

q_{bv} = volumetric sediment discharge per unit width

q_{f0} = flow rate per unit width in floodplain at normal depth (unconstricted)

q_{f0c} = critical flow rate per unit width in floodplain at normal depth (unconstricted)

q_{f1} = flow rate per unit width in the approach floodplain (constricted)

q_{f2} = flow rate per unit width in the contracted floodplain

q_{m1} = flow rate per unit width in the approach main channel at the beginning of scour

q_{m2} = flow rate per unit width in the contracted section in the main channel

Q = total discharge in compound channel

Q_{f1} = discharge in the approach floodplain

Q_{m0} = discharge in main channel for uniform flow in a compound channel

Q_{m1} = discharge in the approach main channel

Q_{m2} = discharge in the contracted main channel

Q_{obst1} = obstructed discharge in the approach section

R = channel hydraulic radius

SG = specific gravity of sediment

t = time since beginning of scour

t_e = time to reach equilibrium scour depth

u_{*1} = approach value of shear velocity = $(\tau_1/\rho)^{0.5}$

u_{*c} = critical value of shear velocity = $(\tau_c/\rho)^{0.5}$

V_{ab} = maximum resultant velocity near the upstream corner of the abutment face

V_c = critical velocity for initiation of motion

V_{f0} = floodplain velocity at normal depth

V_{f0c} = critical velocity for unconstricted floodplain depth of uniform flow

V_{f1} = bridge approach velocity in floodplain of compound channel

V_{f2} = floodplain velocity in the contracted floodplain after scour

V_{m0c} = critical velocity for the unconstricted depth in the main channel

V_{m1} = bridge approach velocity in main channel of compound channel

V_{m2c} = critical velocity in the main channel at the contracted section at equilibrium scour

V_x = mean contraction velocity in GKY scour formula

V_R = resultant velocity adjacent to tip of abutment in GKY scour formula

V_1 = bridge approach velocity in rectangular channel

y = generic symbol for depth of flow in main channel or floodplain for calculating critical velocity

y_{ab} = depth near the upstream corner of the abutment face

y_{f0} = normal depth in floodplain

y_{f1} = bridge approach depth in floodplain of compound channel, including backwater

y_{ftw} = tailwater depth in floodplain

y_{m0} = normal depth in main channel

y_{m1} = bridge approach depth in main channel of compound channel, including backwater

y_{m2} = depth of flow in main channel in contracted section at equilibrium scour

y_{sc} = total depth of flow after scour, including the contraction scour depth only (Maryland formula)

y_1 = bridge approach depth in rectangular channel

y_2 = total flow depth in contracted section after scour (Maryland formula)

y' = vertical coordinate measuring distance above the main-channel bed

y'/y_m = ratio of distance above the main-channel bed to depth of the flow in the main channel

γ = specific weight of fluid

γ_s = specific weight of sediment

μ_f = general discharge per unit width contraction ratio for the floodplain

Φ = ratio of boundary shear force to streamwise component of weight

φ = angle of repose of sediment

ρ = fluid density

ρ_s = sediment density

σ = ratio of Manning's n in the main channel with compound-channel flow to the value for the flow in the main channel alone

σ_g = geometric standard deviation of sediment grain size distribution

τ_c = critical shear stress for initiation of sediment motion

τ_0 = average bed shear stress in uniform flow

τ_1 = bed shear stress in approach flow

τ_* = Shields' parameter, $\tau/[(\gamma_s - \gamma)\, d_{50}]$

τ_{*c} = critical value of Shields' parameter, $\tau_c/[(\gamma_s - \gamma)\, d_{50}]$

τ_{*1} = Shields' parameter in approach flow section

τ_{*2} = Shields' parameter in contracted flow section

ξ = independent dimensionless ratio in proposed scour formula = q_{f1}/Mq_{f0c}

CHAPTER 1. INTRODUCTION

BACKGROUND

Recent bridge failures caused by local scour around piers and abutments have prompted a need for better technical information on scour prediction and scour-protection measures.[1-3] In 1987, the I-90 bridge over Schoharie Creek near Albany, NY, failed because of local scour around the pier foundations, resulting in the loss of 10 lives and millions of dollars for bridge repair/replacement.[2,4] During the 1993 upper Mississippi River basin flooding, more than 2400 bridge crossings were damaged.[5] In 1994, tropical storm Alberto caused numerous bridge failures as a result of the 100-year flood stages being exceeded at many locations along the Flint and Ocmulgee Rivers in central and southwest Georgia.[6]

There are more than 480,000 bridges over water in the United States.[4] As demonstrated in the past decade, the potential for loss of life and serious disruption of a local economy in the event of a bridge foundation failure caused by an extreme flood is very significant. As a result, a comprehensive effort has been undertaken by the Federal Highway Administration (FHWA) to require all States to evaluate highway bridges for scour potential.[7] Approximately 17,000 bridges have been identified by State departments of transportation (DOTs) as "scour critical" with the potential for scour-related failure of the foundations as a result of a flood disaster. Several technical publications have been developed by FHWA, including Hydraulic Engineering Circular Number 18 (HEC-18),[8] to provide guidance to the engineer on evaluating scour problems at bridges. Unfortunately, the alteration of flow patterns by bridge crossings and the concomitant scour process are quite complex and have defied numerical analysis for the most part. The scouring is the result of flow separation around a bridge pier or abutment and the formation of three-dimensional, periodic horseshoe and wake vortexes that interact with a movable sediment bed. In this situation, the engineer has been forced to rely on laboratory results that are numerous and sometimes conflicting because of idealized laboratory conditions that have been used in the past. In particular, the task of predicting local scour around bridge abutments remains problematic, with many engineers not trusting the current empirical formulas given in HEC-18 that are based on laboratory experiments in rectangular flumes.[9]

Previous laboratory experiments on abutment scour have emphasized the abutment length in a rectangular channel as one of the primary variables affecting scour. In an actual river consisting of a main channel and adjacent floodplains, an abutment terminating in the floodplain is not subject to the idealized, uniform approach velocity distribution obtained in previous laboratory experiments in rectangular flumes.[10] Instead, the scour is a function of the redistribution of flow between the main channel and the floodplain as flow through the bridge opening occurs. In other words, abutment length is certainly important; however, the same abutment length may result in different scour depths depending on the approach flow distribution in the compound channel and its redistribution as it flows through the contracted opening.[11]

Currently, FHWA recommends the prediction of abutment scour with a regression equation developed by Froehlich[12] that is based entirely on results from experimental investigators using rectangular laboratory channels. Laursen[13] has developed an equation for clear-water abutment scour that is based on contraction hydraulics, but that relies directly on abutment length.

Melville[14] has proposed a methodology for predicting maximum abutment scour that also depends on abutment length for short- and intermediate-length abutments, and it does not include the effects of overbank flows or of flow distribution in compound channels. However, Melville[15] has also considered the case of compound channels under the condition that the abutment extends into the main channel rather than terminating on the floodplain. This case of encroachment into the main channel itself would be less common in practice than the abutment terminating on the floodplain or at the edge of the main channel.

The concept of flow distribution in a compound channel depends on the interaction between main-channel flow and floodplain flow at the imaginary interface between the two where vortexes and momentum exchange occur. The net result is that less discharge occurs in the compound channel than would be expected from adding the separate main-channel and floodplain flows that would occur without interaction. The research by Sturm and Sadiq,[16] for example, suggests methods by which predictions of flow distribution between the main channel and the floodplain can be improved in the case of roughened floodplains. Wormleaton and Merrett[17] and Myers and Lyness[18] have also proposed techniques for predicting flow distribution in compound channels. A more detailed review of the literature on compound-channel hydraulics can be found in chapter 2 of this report.

Sturm and Janjua[19-20] have proposed a discharge contraction ratio as a better measure of the effect of abutment length, and the flow redistribution that it causes, on abutment scour. The discharge contraction ratio is a function of abutment length and compound-channel geometry and roughness. It can be obtained from the output of the water-surface profile program, WSPRO.[21] The research reported herein, however, attempts to clarify the influence of bridge backwater on the flow redistribution and to improve the WSPRO methodology for computation of flow redistribution by incorporating more recent research results on compound-channel hydraulics.[22]

The abutment scour experiments by Sturm and Janjua[20] and Sturm and Sadiq[23] used a single, uniform sediment size of 3.3 millimeters (mm). The effect of sediment size on the equilibrium scour depth has been incorporated by including, as an independent variable, the ratio of the approach velocity in the floodplain to the critical velocity for the initiation of motion, which depends on sediment size. Experiments with three different sediment sizes at varying discharges and abutment lengths for a vertical-wall abutment were conducted in the present research to verify this method for quantifying the influence of sediment size on equilibrium scour depth. In all cases, the abutment length was large relative to the sediment size in order to remove any effects of energy dissipation caused by large, uniform sediment sizes in the bottom of the scour hole relative to a short abutment length.[24]

The effect of abutment shape was considered in this research by conducting experiments in a compound channel on vertical-wall, wingwall, and spill-through abutment shapes. A single sediment size of 3.3 mm was used for this series of experiments, and the discharge and abutment lengths were varied. The abutment shape effect has been shown to be insignificant for long abutments in rectangular channels[15]; however, this behavior has not been verified for long abutments in compound channels. Currently, FHWA procedures assume a reduction in scour for a spill-through abutment of 55 percent in comparison to the vertical-wall abutment.[8]

Additional experiments were also conducted on abutment lengths that approached the bank of the main channel for both the spill-through and vertical-wall abutments. The purpose of these experiments was to test the methodology developed for abutments that terminated on the floodplain for the more complicated three-dimensional flow field that occurs when the contracted flow joins the main channel near the abutment face.

The live-bed scour case was considered analytically for the condition of sediment transport in the main channel with no sediment movement in the floodplain. This case would be of interest for the abutment located at or near the bank of the main channel with the scour hole occurring at least partially in the main channel rather than on the floodplain alone. Although experiments for this case were attempted, they were not successful because of the limitations of the present compound-channel geometry in the flume.

Finally, a brief implementation procedure for the proposed methodology was developed and applied to a hypothetical example. It was also tested on two field cases of scour in Minnesota using data measured by the U.S. Geological Survey (USGS) in 1997.

RESEARCH OBJECTIVES

The objective of the proposed research is to develop better predictive equations for assessing the vulnerability of existing bridges to abutment scour for cases in which the abutment is located anywhere in the floodplain up to the bank of the main channel in a compound-channel geometry. Of primary interest is the effect of compound-channel hydraulics[16] on the redistribution of the main-channel and floodplain flows as the flow accelerates around the end of an abutment of varying shape. In addition, the effect of sediment size on abutment scour needs to be clarified as well as the effect of live-bed versus clear-water scour.

The experimental research reported herein differs from most of the previous experimental studies of abutment scour that have not had a compound channel as the approach channel and that have not considered the effect of very long abutments in wide, shallow floodplain flow. The results of the experimental research have been used to develop a prediction equation for clear-water abutment scour that was tested on limited field data and was compared to the results of the other scour-prediction methods.

The specific research objectives are:

- Investigate the effects of flow distribution, as affected by abutment length, on clear-water abutment scour in a compound channel for abutment lengths that terminate on the floodplain as well as encroach on the bank of the main channel.

- Quantify the effects of floodplain sediment size on abutment scour.

- Explore the influence of abutment shapes, including wingwall, vertical-wall, and spill-through shapes, on equilibrium scour-hole depth and scour-hole form.

- Determine the relative importance of the live-bed scour case compared to the clear-water case when there is sediment transport in the main channel and the abutment encroaches on the main channel.

- Combine the experimental results into a methodology for assessing field abutment scour and test it on available field data.

This report provides a brief review of the literature on compound-channel hydraulics, and on clear-water and live-bed abutment scour in chapter 2. The experimental investigation is described in detail in chapter 3. Chapter 4 contains an analysis of the experimental results and a proposed abutment scour-prediction equation that addresses the effects of the alteration of the flow distribution by both short and long abutments, sediment size, abutment shape, time development, and live-bed conditions. A procedure for implementing the research results for the purpose of identifying scour-susceptible bridges in the field is then suggested in chapter 4 along with an example of a field application. The proposed procedure is tested with limited field data. Conclusions and recommendations are given in chapter 5.

CHAPTER 2. LITERATURE REVIEW

In this chapter, the literature on compound-channel hydraulics and both clear-water and live-bed bridge abutment scour is reviewed. The connection between these two topics is an important contribution of this research, so a brief summary of previous research done on each topic is given.

COMPOUND-CHANNEL HYDRAULICS

A compound channel consists of a main channel, which carries base flow and frequently occurring runoff up to bank-full flow conditions, and a floodplain on one or both sides that carries overbank flow during times of flooding. Both channel roughness and the depth of the flow, and hence flow-channel geometry, can be drastically different in the main channel and the floodplains. In general, floodplain flows are relatively shallow, with slow-moving flow adjacent to faster-moving flow in the main channel, which results in a complex interaction that includes momentum transfer between the main-channel and floodplain flows. This phenomenon is more pronounced in the immediate interface region between the main channel and the floodplain, where there exists a strong transverse gradient of the longitudinal velocity. Because of the velocity gradient and anisotropy of the turbulence, there are vortexes rotating about both the vertical and horizontal axes along the main channel/floodplain interface.[25-27] These vortexes are responsible for the transfer of water mass, momentum, and species concentration from the main-channel flow into the floodplain flow. The result is that for a given stage, the total flow in the compound channel is less than what would be calculated as the sum of the flows in the main channel and the floodplain assuming no interaction.[25]

Several attempts have been made at quantifying the momentum transfer at the main channel/floodplain interface using the concepts of imaginary interfaces that are included or excluded as the wetted perimeter and defined at varying locations with or without the consideration of an apparent shear stress acting on the interface (see references 17 and 28 through 31). The resulting distribution of the discharge between the main channel and the floodplain caused by the interaction at the interface must be correctly predicted in water-surface profile computations and in calculations of approach floodplain velocities for the prediction of abutment scour.

The current version of WSPRO (as well as the Hydrologic Engineering Center River Analysis System (HECRAS) divides the compound channel into subsections using a vertical interface between the main channel and the floodplain, but neglects any contribution of the interface to the wetted perimeter of the subsections. In effect, the interaction between the main channel and the floodplain is neglected because this procedure is equivalent to assuming no shear stress at the interface. Wright and Carstens[28] proposed that the interface be included in the wetted perimeter of the main channel, and that a shear force equal to the mean boundary shear stress in the main channel be applied to the floodplain interface. Yen and Overton,[29] on the other hand, suggested the idea of choosing an interface on which shear stress was, in fact, nearly zero. This led to several methods of choosing an interface, including a diagonal interface from the top of the main-channel bank to the channel centerline at the free surface, and a horizontal interface from bank to bank of the main channel. Wormleaton and Hadjipanos[32] compared the accuracy of the vertical, diagonal, and horizontal interfaces in predicting the separate main-channel and floodplain discharges measured in an experimental 1.21-meter (m) flume, having a fixed ratio of floodplain width to main-channel half-width of 3.2.[30] The wetted perimeter of the interface was either fully included or fully excluded in the calculation of the hydraulic radius of the main channel. The results showed that even though a particular choice of interface might provide a satisfactory estimate of total channel discharge, nearly all of the choices tended to overpredict the separate main-channel discharge and underpredict the floodplain discharge. It was further shown that these errors were magnified in the calculation of the kinetic energy flux correction coefficients used in water-surface profile computations.

In a modification of the earlier method, Wormleaton and Merrett[17] applied a correction factor called the Φ index to the main-channel and floodplain discharges calculated by a particular choice of interface (vertical, diagonal, or horizontal), which was either included or excluded from the wetted perimeter. The Φ index was defined as the ratio of the boundary shear force to the streamwise component of fluid weight[33] as a measure of apparent shear force. If the Φ index is less than unity on the main-channel interface, for example, then the apparent shear force resulting from the main channel/floodplain interaction resists the fluid motion in the main channel. This modified method was applied to the experimental results from the very large compound channel at Hydraulics Research, Wallingford Ltd. (Wallingford, Oxfordshire, U.K.). The channel is 56 m long by 10 m wide, with a total flow capacity of 1.1 cubic meters per second (m³/s). In the experiments, the ratio of the floodplain width to the main-channel half-width varied from 1 to 5.5, and the ratio of the relative floodplain depth to the main-channel depth varied from 0.05 to 0.50. The calculated main-channel and floodplain discharges, when multiplied by the square root of the Φ index for each subsection, showed considerable improvement when compared to the measured discharges, regardless of the choice of interface. The difficulty of the method is in the prediction of the Φ index. A regression equation was proposed for this purpose, with the Φ index given as a function of the velocity difference between the main channel and the floodplain, the floodplain depth, and the floodplain width. The regression equation is limited to the range of experimental variables observed in the laboratory.

Ackers[34] has also proposed a discharge calculation method for compound channels using the Wallingford data. He suggested a discharge adjustment factor that depends on *coherence*, which was defined as the ratio of the full-channel conveyance (with the channel treated as a single unit with perimeter weighting of boundary friction factors) to the total conveyance calculated by summing the subsection conveyances. Four different zones were defined as a function of the relative depth (the ratio of the floodplain to the total depth), with a different empirical equation for the discharge adjustment for each zone.

Myers and Lyness[18] have proposed a two-step method for predicting the distribution of flow between the main channel and the floodplains for a compound channel in overbank flow. First, they propose a power relationship between the ratio of the total discharge to the bank-full discharge and the ratio of the total depth to the bank-full depth. They show that from both laboratory and field data that such a relationship is independent of bed slope and scale, and is dependent only on geometry for similar roughnesses in the main channel and the floodplain. Second, they divide the total discharge obtained from the first step into the main-channel and floodplain components, assuming that the ratio of main channel-to-floodplain discharge is an inverse power function of the ratio of the total depth to the floodplain depth. In this case, the relationship was not shown to be universal for different geometry, scale, and roughness. This approach is, in reality, a conveyance weighting method that does not account explicitly for the interaction between the main channel and the floodplain.

The variation of Manning's n with depth further complicates the problem of water-surface profile computations and velocity predictions in compound channels.[16,35] In addition, the existence of multiple critical depths can lead to difficulties in both computing and interpreting water-surface profiles. Blalock and Sturm[36] and Chaudhry and Bhallamudi[37] have suggested the use of a compound-channel Froude number in computational procedures for determining multiple critical depths in compound channels. Blalock and Sturm[36] used the energy equation, while Chaudhry and Bhallamudi[37] employed the momentum equation to define a compound-channel Froude number. Blalock and Sturm[38] showed from their experimental results that the energy and momentum approaches resulted in the same values for critical depth. Yuen and Knight[39] have confirmed from their experimental results that the compound-channel Froude number suggested by Blalock and Sturm[36] gives values for critical depth that are reasonably close to the measured values. Sturm and Sadiq[40] have demonstrated the usefulness of their compound-channel Froude number for computing and interpreting water-surface profiles in compound channels having two values for critical depth: one in overbank flow and one in main-channel flow alone.

Numerical analysis of flow characteristics in compound open channels, particularly for the case of an obstruction, such as a bridge abutment on the floodplain, have received rather limited attention in comparison to experimental studies. Most numerical models have been developed for the parabolic flow situation, which has a predominantly longitudinal flow direction with no reverse flow (e.g., uniform or gradually varied flow). There are, however, a few studies applied to an elliptical flow situation (with flow separation and recirculation allowed), but these are applicable only to a simple rectangular channel. The effect of compound-channel hydraulics on flow characteristics in the region close to obstructions is in need of further investigation.

One- and two-dimensional numerical models have been used to compute complex flow fields, such as those in compound open-channel flow. Some of the two-dimensional models[41-42] are applicable to the boundary-layer type of flow, in which the flow can be described by a set of differential equations that are parabolic in the longitudinal (flow) direction.[43] The flow situation in a compound open channel can become even more complex if structures, such as bridge piers or abutments, are placed in the floodplain. In this situation, at least for the regions near the structure, the parabolic flow assumption becomes invalid. This is because there is no longer any predominant flow direction in the region close to the structure. The flow will separate downstream from the structure and a recirculating region with a reverse flow and an adverse pressure gradient will be formed that violates the parabolic flow assumption. In addition, the water-surface elevation no longer varies only in the streamwise coordinate. Because of the presence of the obstruction, there exists a rapid variation in water depth near a bridge abutment, both in the longitudinal and transverse directions. The water-surface elevation upstream from the abutment increases and forms a backwater profile. These flow characteristics are associated with an elliptical flow field that has additional complexity because of the intercoupling of the velocity and pressure (water depth) fields.

Three-dimensional models have been developed and applied primarily to the case of uniform flow in a compound channel.[44-46] These models allow the simulation of turbulence-driven secondary motion in the transverse plane of a compound open channel, as well as the Reynolds' turbulent shear stresses. Experimental measurements by Tominaga and Nezu[27] using a fiber-optic laser Doppler anemometer showed a pair of longitudinal vortexes at the main channel/floodplain interface, with an inclined secondary current directed from the intersection of the main channel and the floodplain bed toward the free surface. The numerical model of Naot, et al.,[46] was able to reproduce this behavior. The numerical results of Krishnappen and Lau[44] using an algebraic stress model demonstrated good agreement with measured divisions of flow between the main channel and the floodplain using data from Knight and Demetriou[31] and Wormleaton, et al.[30] Pezzinga's nonlinear k-ε turbulence model[47] was used to compare the effects of various main channel/floodplain subdivisions (vertical, horizontal, diagonal, and bisector) on computed subsection discharges. The diagonal interface and the interface formed by the bisector of the corner angle between the main channel and the floodplain gave the best results for both the discharge distribution and the kinetic energy correction coefficients.

In this research, the velocity field is sought in the bridge approach section and near the face of a bridge abutment that terminates on the floodplain where the width-to-depth ratio of the flow is large and the vertical mixing is strong. Under these conditions, the velocity field can be computed adequately from the depth-averaged equations of motion, including the regions of adverse pressure gradient and flow separation, especially if a k-ε turbulence closure model is used.[42, 48-53] Biglari[54] and Biglari and Sturm[55] applied a depth-averaged k-ε turbulence model in elliptical form to the problem of flow around a bridge abutment on a floodplain.

ABUTMENT SCOUR

Early experimental research on scour around abutments was, in some cases, motivated by a desire to predict local scour around spur dikes; however, the results of these studies have been applied to the problem of abutment scour. Ahmed[56] proposed a scour-depth relationship for spur dikes that used the "flow intensity," or flow rate per unit width in the contracted section, as the independent variable in agreement with previous regime formulas. Laursen and Toch[57] argued that in live-bed scour around bridge piers and abutments, the approach flow velocity has no effect on equilibrium scour depth because increases in velocity not only increase the sediment transport into the scour hole from upstream, but also increase the strength of the vortex responsible for transporting sediment out of the scour hole. This argument applies in cases where there is appreciable sediment in bed-load transport. If the sediments in transport are fine and mostly carried as suspended loads, the assumptions of live-bed scour behavior may not be appropriate.

Garde, et al.,[58] studied experimentally the scour around spur dikes in a rectangular channel and related the nondimensional ratio of scour depth to approach depth, d_s/y_1, to the approach Froude number, F_1, and to the geometric contraction ratio, m, which is defined as the ratio of the width of the contracted opening to the approach channel width. In addition, the coefficient of proportionality and the exponent in the Froude number were found to depend on sediment size. Some sediment transport occurred in the approach flow for these experiments.

Liu, et al.,[59] considered the scour around bridge constrictions caused by abutment models in 1.2-m- (4-foot (ft)-) wide and 2.4-m- (8-ft-) wide flumes. Their experimental results for live-bed scour indicated that the ratio of the abutment length to the normal depth, L_a/y_0, and the uniform-flow Froude number were the most important influences on the dimensionless scour depth. The normal depth was determined with sediment in equilibrium transport before the abutment was placed in the flume. They proposed an equation for equilibrium live-bed scour depth d_s:

$$\frac{d_s}{y_0} = 2.15 \left(\frac{L_a}{y_0}\right)^{0.4} \mathbf{F}_0^{1/3}$$
(1)

where L_a = abutment length (vertical wall), y_0 = normal depth of flow, and F_0 = Froude number of uniform flow. The experimental values of L_a/y_0 varied from approximately 1 to 10, and the Froude numbers varied from 0.3 to 1.2. In a separate series of experiments, clear-water scour was studied by pre-forming the scour hole and determining the flow conditions necessary to just initiate sediment motion in the bottom of the scour hole. In this case, the dimensionless clear-water scour depth was found to be directly proportional to F_0/m, where F_0 = uniform-flow Froude number and m = geometric contraction ratio defined by the ratio of the constricted channel width to the approach channel width. The coefficient of proportionality was approximately 12.5.

Gill[60] argued from his experimental results on the scour of sand beds around spur dikes that the distinction between clear-water and live-bed scour is unimportant for the design determination of maximum scour depth. He proposed that the maximum scour depth be based on the geometric contraction ratio, m, and on the ratio of the sediment size to the flow depth based on both clear-water and live-bed scour experiments having a duration of 6 hours. His proposed equation is:

$$\frac{d_s}{y_0} + 1 = 8.38 \left(\frac{d_{50}}{y_0}\right)^{1/4} \left(\frac{1}{m}\right)^{6/7}$$
(2)

where y_0 = initial uniform-flow depth, d_{50} = median sediment grain size, and m = geometric contraction ratio given by the ratio of the contracted width to the full channel width.

Laursen[61] developed scour-depth relationships for bridge abutments that were based on treating the abutment as a limiting case of scour through a long flow constriction. Live-bed scour was considered to be a function only of the ratio of the abutment length to the approach flow depth, L_a/y_1, and the ratio of the discharge per unit width in the overbank flow region to the discharge per unit width in the scour region. The scour region was assumed to have a constant width of 2.75 times the scour depth. In a subsequent analysis of relief-bridge scour,[13] which was considered to be a case of clear-water scour, the same approach was taken in relating the abutment scour to that which would take place in a long constriction. The contracted width was assumed to be equal to a scour-hole width of 2.75 times the scour-hole depth. This assumption resulted in an implicit relationship for scour depth:

$$\frac{L_a}{y_1} = 2.75 \frac{d_s}{y_1} \left[\frac{(\frac{1}{11.5} \frac{d_s}{y_1} + 1)^{\frac{7}{6}}}{(\frac{\tau_1}{\tau_c})^{\frac{1}{2}}} - 1 \right] \tag{3}$$

where L_a = abutment length, y_1 = approach flow depth, d_s = maximum scour depth, τ_1 = bed shear stress in the approach flow, and τ_c = critical shear stress for initiation of sediment motion.

In a comprehensive experimental study of the effect of flow depth on clear-water scour around abutments, Tey[62] held constant the ratio of approach shear stress to critical shear stress, τ_1/τ_c, at a value of 0.90, while varying the flow depth and the abutment length and shape. The results showed an increasing scour depth with increasing flow depth, but at a decreasing rate as the depth became larger. The length of the abutment obstructing the approach flow and the abutment shape were also found to influence the scour depth. Longer abutments and blunter abutment shapes caused deeper scour holes.

Froehlich[12] completed a regression analysis of 164 laboratory experiments from 11 separate sources on clear-water scour around abutments or spur dikes. His proposed regression equation is:

$$\frac{d_s}{y_1} = 0.78 \ K_1 \ K_2 \ \left(\frac{L_a}{y_1}\right)^{0.63} \ F_1^{1.16} \ \left(\frac{y_1}{d_{50}}\right)^{0.43} \ \sigma_g^{-1.87} + 1 \tag{4}$$

where d_s = scour depth; y_1 = approach flow depth, K_1 = geometric shape factor for abutment and embankment, K_2 = embankment skewness factor, L_a = abutment length, F_1 = approach Froude number, d_{50} = median sediment grain size, and σ_g = geometric standard deviation of the sediment size distribution. Froehlich further proposed that a factor of safety (FS) of 1 should be added to the value of d_s/y_1 obtained from the regression analysis, and it has been included on the right-hand side of equation 4. Currently, HEC-18 recommends the live-bed scour equation obtained by Froehlich's regression analysis of other investigators' results for this case because equation 4 seems to greatly overestimate abutment scour. The live-bed scour equation is:

with FS = 1 included on the right-hand side. In both equations 4 and 5, Froehlich calculated the approach Froude number based on an average velocity and depth in the obstructed area of the approach-flow cross section. While this worked well for the experimental results that he used, which were for rectangular channels, it is not clear what the representative approach velocity and depth should be in natural channels subject to overbank flow.

Melville[14] summarized a large number of experimental results on clear-water abutment scour

$$\frac{d_s}{y_1} = 2.27 \ K_1 \ K_2 \ \left[\frac{L_a}{y_1}\right]^{0.43} \ \frac{F_1^{0.61}}{10} + 1 \tag{5}$$

from rectangular channels and proposed a design method for maximum scour depth that depends on empirical correction factors for flow intensity, abutment shape, alignment, and length. He classified abutments as short ($L_a/y_1 < 1$) or long ($L_a/y_1 > 25$), and suggested that the maximum clear-water scour depth was $2L_a$ for the former case and $10y_1$ for the latter case. For intermediate abutment lengths, the equilibrium clear-water scour depth was given as:

$$d_s = 2 \; K_I \; K_s^* \; K_\theta^* \; (y_1 \; L_a)^{0.5} \tag{6}$$

where K_I = flow-intensity factor (V_1/V_c), V_1 = approach velocity, V_c = critical velocity for initiation of sediment motion, K_s^* = abutment shape factor, K_θ^* = abutment alignment factor, y_1 = approach flow depth, and L_a = abutment length.

Subsequently, Melville[24] suggested that the same methodology could be applied to both bridge piers and abutments, albeit with slightly different equations. He further showed that sediment size effects appear only in the flow-intensity factor for clear-water scour as long as $L_a/d_{50} > 25$. Abutment shape effects were reported to be important only for shorter abutments (i.e., $K_s^* = 1.0$ for vertical-wall abutments; 0.75 for wingwall abutments; and from 0.45 to 0.60 for spill-through abutments), depending on the sideslope, but only if $L_a/y_1 \leq 10$. For long abutments ($L_a/y_1 \geq 25$), $K_s^* = 1.0$ with a linear relationship between K_s^* and L_a/y_1 for the intermediate range of $L_a/y_1 = 10$ to 25. It must be emphasized that all of Melville's integrated abutment/pier results[24] are considered to be for abutments that are significantly shorter than the floodplain width itself so that flow contraction effects are not important.

Live-bed abutment scour results were also summarized by Melville[24] for *short abutments* using Dongol's data.[63] Under clear-water conditions ($V_1 < V_c$), the scour depth increased to a maximum at $V_1/V_c = 1$. For live-bed conditions ($V_1 > V_c$), the scour depth decreased slightly as V_1 increased, but then increased again to a value equal to the maximum clear-water scour. Piers showed the same behavior under live-bed conditions.

Melville and Ettema[64] and Melville[15] reported research results on abutment scour in a compound channel, but for the case of an abutment terminating in the main channel rather than on the floodplain. It was proposed that the scour depth in this instance could be calculated as the scour in an equivalent rectangular channel of the same width and with a depth equal to the main-channel depth by multiplying equation 6 by a geometric factor given as:

$$K_G = \sqrt{1 - \frac{B_f}{L_a}\left[1 - \left(\frac{y_{f1}}{y_{m1}}\right)^{5/3} \frac{n_{mc}}{n_{fp}}\right]} \tag{7}$$

where L_a = abutment length extending into main channel, B_f = width of floodplain, y_{f1} = depth of flow in floodplain, y_{m1} = depth of flow in main channel, n_{mc} = Manning's n in main channel, and n_{fp} = Manning's n in floodplain.

Sturm and Janjua[19-20] conducted a series of experiments in a compound channel consisting of a main channel and a floodplain with the abutment terminating in the floodplain, and showed that the approach flow distribution and its readjustment by the abutment in the contracted section are important factors that should be included in equations for predicting scour depth in natural channels. On the basis of a dimensional analysis and the application of Laursen's[13] analysis of relief bridge scour in a long contraction to compound channels, they proposed a relationship given as:

$$\frac{d_s}{y_{f_1}} = 7.70 \left[\frac{V_{f_1}/V_c}{M} - 0.35\right] \tag{8}$$

where d_s = equilibrium scour depth, y_{f_1} = approach floodplain flow depth, V_{f_1} = approach floodplain flow velocity, V_c = critical velocity, and M = discharge contraction ratio defined as the fraction of the total discharge in the bridge approach section over a width determined by extending the bridge opening upstream to the approach section. Melville[15] postulated that these experimental results gave smaller scour depths than his equation for intermediate-length abutments because the experiments had not reached equilibrium; however, subsequent experimental results given by Sturm and Sadiq[23] for a different compound-channel geometry, but at much longer scour durations, were similar to those of Sturm and Janjua.[20] Young, et al.,[65] developed a regression equation for clear-water as well as live-bed abutment scour using the calculated contraction scour as a nondimensionalizing parameter for the abutment scour depth. More recently, Young, et al.,[66] suggested an abutment scour equation:

$$(y_1 + d_s) = 1.37 \left[\frac{n^2}{\tau_{*c}(SG-1)d_{50}}\right]^{3/7} (y_1 V_R)^{6/7} \tag{9}$$

where y_1 = approach depth in meters, d_s = scour depth in meters, n = Manning's resistance factor, τ_{*c} = critical value of Shields' parameter, SG = specific gravity of sediment, d_{50} = median particle size in meters, and V_R = resultant velocity adjacent to the tip of the abutment in meters per second. The resultant velocity is calculated from $V_x/\cos\theta$, where V_x = mean contraction velocity from continuity and $\theta = 69.85 (a/A)^{0.2425}$ (with a coefficient of determination of $r^2 = 0.54$), where a = blocked flow area by the abutment and A = total unobstructed flow area including the main channel to the median flow bisector. Equation 9 is not dimensionally homogeneous and is meant for the International System of Units (SI), known as the metric system. It was tested on an experiment by Lim[67] with a very short abutment ($L_a/y_1 = 1$) and showed good agreement for this case.

Chang[68] has applied Laursen's long contraction theory to both clear-water and live-bed scour. He suggested a velocity adjustment factor, k_v, to account for the nonuniform velocity distribution in the contracted section, and a spiral-flow adjustment factor, k_f, at the abutment toe that depends on the approach Froude number. The value of k_v was based on potential flow theory, and k_f was determined from a collection of abutment scour experiments in rectangular laboratory flumes.[69] The resulting scour equation was:

$$\frac{y_2}{y_1} = k_f \left(\frac{k_v \, q_2}{q_1}\right)^\theta \tag{10}$$

where y_2 = flow depth in contracted section after scour, y_1 = approach flow depth, q_1 = flow rate per unit width in approach section, q_2 = flow rate per unit width in contracted section, and θ = 0.857 for clear-water scour. The value of $k_v = 0.8 \, (q_1/q_2)^{1.5} + 1$ and $k_f = 0.1 + 4.5F_1$ for clear-water scour, while $k_f = 0.35 + 3.2F_1$ for live-bed scour. The approach Froude number $F_1 = V_1/(gy_1)^{0.5}$. Equation 10 does not include the effect of sediment size on clear-water abutment scour. It has since been modified[70] to the form:

$$y_2 = k_f \, (k_v)^{0.857} \, y_{sc} \tag{11}$$

where y_2 = total depth of flow at the abutment including scour depth, y_{sc} = total depth of flow including the contraction scour depth only, and k_f and k_v are unchanged from the previous formulation. The value of y_{sc} is calculated from q_2/V_c, where q_2 = unit discharge at the contraction and V_c = critical velocity obtained from the expressions given by Neill.[71] The evaluation of q_2 is unclear for the case of the contracted section having a compound section with a variable q_2 across the cross section.

Lim[67] has derived an equation for predicting clear-water abutment scour:

$$\frac{d_s}{y_1} = K_s^* \, (0.9 \, X - 2) \tag{12}$$

where d_s = scour depth, y_1 = approach flow depth, K_s^* = abutment shape factor, and X is expressed as:

$$X = \frac{F_0^{0.75} \left(\frac{d_{50}}{y_1}\right)^{0.25}}{\tau_{*c}^{0.375}} \left[0.9 \left(\frac{L_a}{y_1}\right)^{0.5} + 1 \right] \tag{13}$$

where F_0 = approach value of the densimetric grain Froude number, τ_{*c} = critical value of Shields' parameter, d_{50} = median grain size, and L_a = abutment length. Equations 12 and 13 are derived on the basis of satisfying continuity before and after scour, evaluating the velocity before and after scour in the contracted section from a power law with an exponent of 1/3, and using an

13

expression for the shear velocity in the contracted expression proposed by Rajaratnam and Nwachuku.[72] The latter expression is limited to values of $L_a/y_1 \leq 1$. Lim tested equations 12 and 13 on his own abutment scour data as well as on data from Dongol,[63] Rajaratnam,[73] and Liu, et al.,[59] that were, for the most part, limited to very short abutments with $L_a/y_1 \leq 1$.

Lim[74] has also proposed an abutment scour-prediction equation for live-bed scour. He assumed that the sediment transport in a strip of the approach section, with a width equal to the abutment length plus the scour-hole width, is carried completely through the scour-hole width in the contracted section. The scour-hole width is estimated as $d_s/\tan \varphi$, where φ = angle of repose of the bed material. Then, by using a sediment transport relationship similar to that of Meyer-Peter and Muller (Julien[75]), and by making the same assumptions as for his clear-water scour equation, the resulting equation given by Lim for live-bed scour is:

$$(1 + \frac{d_s}{2 y_1})^{4/3} = \frac{1 + 1.2 \sqrt{\dfrac{L_a}{y_1}}}{\sqrt{\dfrac{u_{*c}^2}{u_{*1}^2} + (\dfrac{L_a \tan \varphi}{d_s} + 1)^{2/3} (1 - \dfrac{u_{*c}^2}{u_{*1}^2})}} \tag{14}$$

where u_{*1} = approach value of shear velocity and u_{*c} = critical value of shear velocity. When $u_{*c}/u_{*1} > 1$, the term $[1 - (u_{*c}/u_{*1})^2]$ is taken as equal to zero and equation 14 reduces to the clear-water scour case according to Lim. This equation still suffers from the dependence on abutment length and the limitation on the expression for shear stress for very short abutments as pointed out for the clear-water scour case by Richardson.[76]

An equation in *Highways in the River Environment*[77] was developed from the U.S. Army Corps of Engineers' field data for scour at the end of spur dikes on the Mississippi River. It is recommended in HEC-18 for predicting scour around long abutments with $L_a/y_1 > 25$. The equation is:

$$\frac{d_s}{y_{ab}} = 4 \, \mathbf{F}_{ab}^{0.33} \tag{15}$$

where y_{ab} = depth of flow at the abutment and F_{ab} = Froude number based on the velocity and depth adjacent to the abutment.

TIME DEVELOPMENT OF SCOUR

Within the context of this report, the time to equilibrium of the scour process is important in establishing the necessary duration of the experiments. In addition, as discussed in a previous report by the author,[78] one consequence of developing scour equations from laboratory data for the equilibrium condition is that in the field case for small watersheds, the duration of the design discharge may be considerably shorter than the time to reach equilibrium. The result is an overestimate of field scour for a given design event.

Time development of scour was first studied in Laursen's pioneering research on scour by jets.[79] Laursen argued that clear-water scour is an asymptotic process in which scour depth increases linearly with the logarithm of time. Theoretically, equilibrium would never be reached under these circumstances except at infinity. Thus, as a practical matter, some limiting depth of scour is determined in laboratory experiments during long durations when the rate of change in the scour depth is very small. Most of the research on the time development of scour has been done for situations other than abutment scour (e.g., jet scour, pier scour, or hydraulic structure scour). Carstens[80] has proposed a general method for determining the time development of scour by combining: (1) the sediment continuity equation, (2) a sediment pickup function, and (3) an assumed scour-hole shape that is geometrically similar at all times. He showed that the simultaneous solution of these three equations agreed with Laursen's experimental results for the time development of scour from a horizontal jet, although the experimental data were used to develop the sediment pickup function.

Raudkivi and Ettema[81] showed from their experimental results on the development of pier scour that dimensionless scour depth (d_s/b) was linearly proportional to the logarithm of td_{50}/b^3, where d_s = scour depth, b = cylindrical pier diameter, d_{50} = median sediment grain diameter, and t = time. Yanmaz and Altinbilek[82] applied Carstens' method to the problem of time development of scour around bridge piers. The resulting differential equation was solved numerically to obtain a curve of d_s/b as a function of dimensionless time:

$$\frac{d_s}{b} = f[\frac{t\, d_{50}\, \sqrt{(SG - 1)\,gd_{50}}}{b^2}]$$
(16)

where SG = specific gravity of the sediment. The function was linear with the logarithm of dimensionless time up to a large value of time after which it leveled off in agreement with their experimental results. Kothyari, et al.,[83] proposed a different approach in which the temporal variation of scour around a bridge pier was calculated in successive time steps governed by: (1) the time required for removal of one sediment grain by the primary vortex and (2) the resulting gradual decrease of shear stress in the scour hole as it grew larger.

Chiew and Melville[84] suggested from their experimental results on scour development around bridge piers an empirical relationship for a dimensionless equilibrium time, $t^* = V_1 t_e/b$, as a function of V_1/V_c, where V_1 = approach velocity, V_c = critical velocity, b = pier diameter, and t_e = equilibrium time. The equilibrium time was defined as the time at which the rate of increase of scour depth was less than 5 percent of the pier diameter in 24 hours. The experimental results were then presented as the ratio of scour depth at a given time to equilibrium scour depth (d_{st}/d_s) as a function of the ratio of elapsed time to equilibrium time (t/t_e) for various constant values of V_1/V_c.

A relationship for time development of scour is proposed in this report based on the experimental results for compound channels. It is used to better describe the effect of event duration on the scour depth, and it is incorporated into the proposed scour-prediction procedure. However, it must be used carefully, because scour can be cumulative over many events.

CHAPTER 3. EXPERIMENTAL INVESTIGATION

INTRODUCTION

Experiments were conducted in a 4.2-m-wide by 24.4-m-long flume of a fixed slope in the Hydraulics Laboratory of the School of Civil and Environmental Engineering at the Georgia Institute of Technology. Scour depths were measured as a function of discharge, sediment size, and abutment shape and length for two different compound-channel configurations constructed in the flume at two different fixed slopes. Velocity distributions at the bridge approach section and complete water-surface profiles at the beginning and end of scour were measured. The details of the experiments and some typical results are described in this chapter, and the results are analyzed in chapter 4.

COMPOUND-CHANNEL AND ABUTMENT GEOMETRY

Compound-channel sections were constructed inside the 4.2-m-wide by 24.4-m-long flume for two different series of experiments and are shown in figure 1. Compound channel A was constructed for a series of experiments reported by Sadiq[22] and Sturm and Sadiq.[23] Compound channel A was replaced by compound channel B for use in this study as well as in a separate Georgia DOT study.[78] Compound channel A had a total inside width of 2.13 m and a fixed-bed slope of 0.0050, while compound channel B was 4.21 m wide with a fixed-bed slope of 0.0022 as shown in figure 1.

Steel rails on the top of the flume walls were adjusted to serve as a level track for the instrument carriage, which is driven by an electric motor. A pumping system provided the water supply to the flume from a recirculating sump into which the flume discharges. The flume discharge was measured by calibrated bend meters with an uncertainty in discharge of ±0.0003 cubic meters per second (m^3/s), and the total capacity of the pumping system was approximately 0.20 m^3/s. The tailwater depth in the flume was adjusted with a motor-driven tailgate at the downstream end of the flume. The flume entrance included a head tank and a series of stilling devices to remove the flow entrance effects. The compound-channel section terminated 3.1 m upstream of the tailgate in order to provide a sedimentation tank for the settling of any sediment scoured from the channel. As a result of the length occupied by the head tank and the sedimentation tank, the actual length of the compound channel was 18.3 m.

The main-channel bed of compound channel A consisted of concrete having a longitudinal slope of 0.005 with a tolerance of ±0.001 m from the required elevation at any station along the channel. Fine gravel with a mean diameter of 3.3 mm and a geometric standard deviation of 1.3 was used to provide the roughness on the main-channel bed and walls. Gravel was affixed to the main-channel bed and walls with varnish. Roughness in the floodplain was also provided by gravel of the same size. To prevent movement of the gravel in the floodplain for the uniform-flow experiments, the upper 5-centimeter (cm)-thick layer was stabilized by mixing portland cement in the gravel in a ratio of 1:6 by volume. Water was then added until a reasonable workability was achieved to form a fixed bed consisting predominantly of gravel roughness in the resulting matrix. In the case of compound channel B, a lean concrete mix that used the 3.3-mm gravel as aggregate was poured to a constant slope of 0.0022 with a standard error of

±0.002 m in bed elevation. In both compound channels, the bed surfaces were relatively rough as determined by the protruding grains of gravel.

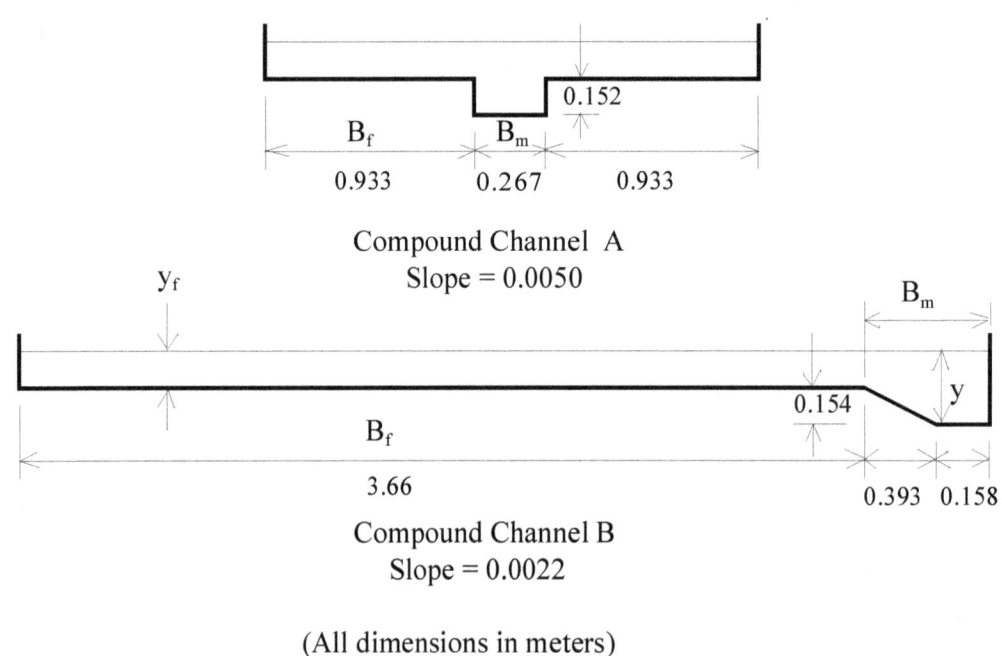

Figure 1. Compound-channel configurations used in scour experiments.

In the case of compound channel A, the bridge abutments were formed by a row of rectangular concrete blocks measuring 0.15 m high by 0.15 m wide (in the flow direction) that were fixed to the floor of the flume with cement mortar at station 9.75 located 9.75 m (32 ft) downstream of the flume entrance. In the text that follows in this report flume station refers to the distance from the flume entrance in meters. The abutments for compound channel B were constructed as a row of concrete blocks poured into custom-built forms to create vertical-wall (VW), spill-through (ST), and wingwall (WW) abutment shapes as shown in figure 2. The sideslopes for the wingwall and spill-through shapes were 2:1 (horizontal:vertical) and the wingwall angle was 30 degrees as shown in figure 2. The centerline of the abutment for compound channel B was also located at station 9.75 as it was for compound channel A. The abutment lengths, L_a, and relative abutment lengths, L_a/B_f, for both compound channels are summarized in table 1.

Table 1. Experimental parameters.

Shape	Compound Channel	Lengths (L_a), m	L_a/B_f	Sediment	d_{50}, mm
Vertical-Wall (VW)	A	0.152, 0.305, 0.457	0.17, 0.33, 0.50	A	3.3
Vertical-Wall (VW)	B	0.80, 1.60, 2.40, 3.23, 3.55, 3.66	0.22, 0.44, 0.66, 0.88, 0.97, 1.0	A, B, C	3.3, 2.7, 1.1
Spill-Through (ST)	B	1.17, 1.97, 2.37, 3.23, 3.55, 3.66	0.32, 0.54, 0.65, 0.88, 0.97, 1.0	A	3.3
Wingwall (WW)	B	1.61, 2.22	0.44, 0.61	A	3.3

SEDIMENT

Three sediments having median sediment grain sizes (d_{50}) of 3.3, 2.7, and 1.1 mm were used in this research as indicated in table 1 where they are referred to as sediments A, B, and C, respectively. The measured size distributions of all three sediments are shown in figure 3. The geometric standard deviation σ_g ($= [d_{84}/d_{16}]^{0.5}$) of sediments A, B, and C was approximately 1.3. These sediments can be considered to be uniform in terms of the size distribution effects on scour.

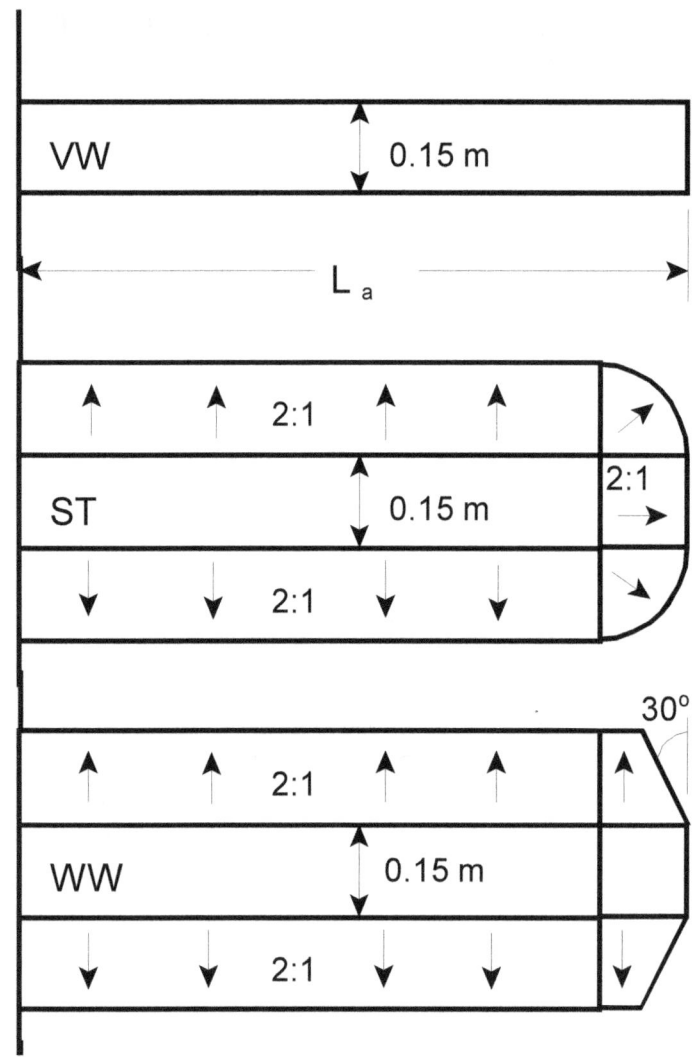

Figure 2. Abutment shapes used in scour experiments.
VW = vertical wall, ST = spill-through, WW = wingwall.

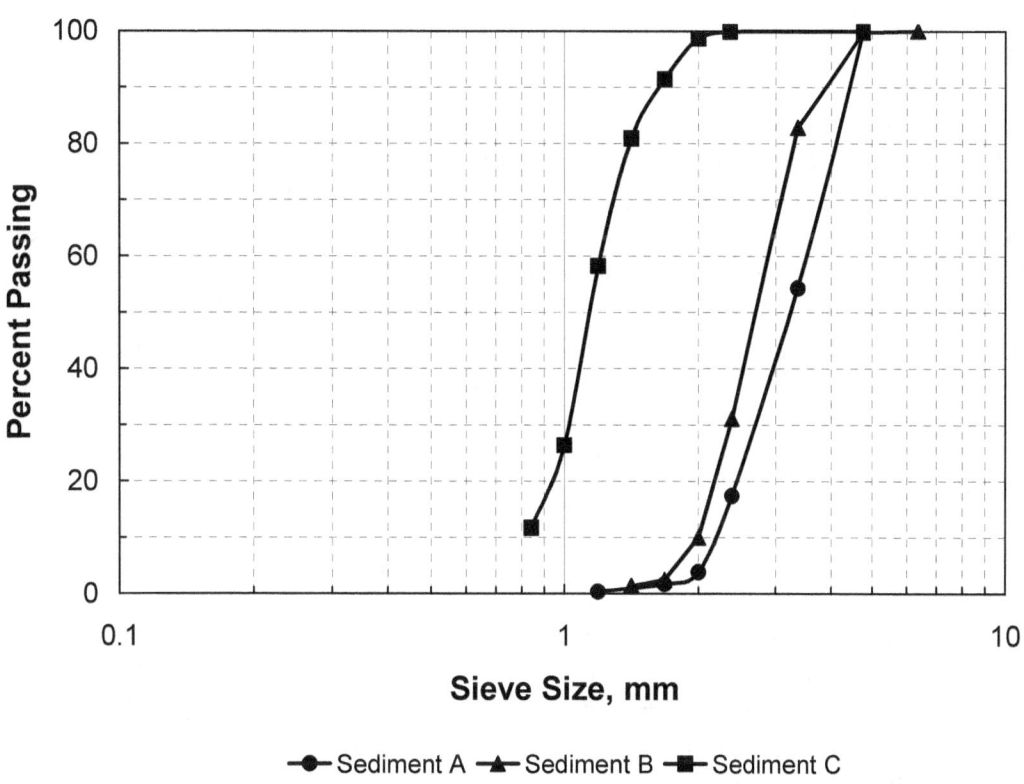

Figure 3. Sediment grain size distributions.

WATER-SURFACE PROFILE AND VELOCITY MEASUREMENTS

Both water-surface profiles and velocities were measured in the fixed-bed compound channels just described, first for uniform flow to determine the bed roughness, and then with the abutments in place to obtain the hydraulic conditions at the beginning of scour. In the uniform-flow experiments, stations for measuring the water-surface elevations were fixed at 1.22-m intervals along the length of the channel. The desired discharge was obtained by adjusting the inlet valve, and the tailgate was raised or lowered to achieve the required M1 or M2 profile with no abutments in place, where M1 and M2 are customary textbook symbols for water surface profiles in open channel flow on a mild slope. M1 profile flow decelerates in the downstream direction; M2 profile flow accelerates in the downstream direction. At each station, the water-surface elevation was measured at three points across the compound channel for compound channel A, and at eight points for compound channel B. The mean of these elevations was then used to determine the mean elevation of the water surface at each station. The normal depths were determined by taking the mean of the flow depths at a point where both M1 and M2 water-surface profiles asymptotically approached one another in the upstream direction, which was generally about 6 to 9 m downstream of the channel entrance. The relative depth ratio (y_{f0}/y_{m0}), where y_{f0} = normal depth in the floodplain and y_{m0} = corresponding normal depth in the main channel (see figure 1), varied from 0.26 to 0.35 for compound channel A, and from 0.13 to 0.32 for compound channel B.

A miniature propeller current meter with a diameter of 1.5 cm was used to measure point velocities averaged over 60 s. In the uniform-flow experiments, station 9.75 was selected for measuring the point velocities across the channel cross section because at this station there were negligible entrance effects, and also because the water surface could be adjusted easily to the normal depth at this station. Depth-averaged velocity distributions were obtained from a point velocity measurement at an elevation above the channel bottom of 0.37 times the flow depth in each vertical profile in the floodplain, and from six to eight point measurements spaced out over each vertical profile in the main channel. Fourteen to 19 vertical profiles were measured at a single cross section to establish the depth-averaged velocity distribution. The resulting measured velocity distributions when integrated over the cross section produced discharges that were within ±4 percent of the discharges measured by the calibrated bend meters in the flume supply pipes for the uniform-flow experimental runs.

The separate discharges in the main channel and the floodplains were determined from the point velocities by integration. This information was combined with the measured normal depths to determine the separate main-channel and floodplain roughnesses in the uniform compound-channel flow from Manning's equation. Normal depth was also determined with flow in the main channel alone in order to determine the equivalent sand-grain roughness of the main channel.

Measurement of water-surface profiles and point velocities was also conducted with the abutments in place in the fixed-bed channel. Away from the abutments, the water-surface elevations were measured at stations 1.22 m apart as in the uniform-flow case. Close to the abutments, water-surface elevations were measured at stations 0.3 m to 0.6 m apart. The depth of the flow at the downstream end of the channel was set at the normal depth for each discharge by adjusting the tailgate. For the longer abutments, the tailwater was set higher than the normal depth in some instances to limit the maximum scour that could occur.

With the abutments in place, and for both a fixed bed at the beginning of scour and a movable bed at the end of scour, detailed point velocity measurements were made across the entire channel cross section at the bridge approach section, which was station 8.5 for compound channel A, and either station 6.7 or station 7.3 for compound channel B, depending on the abutment length. This station was located in the region where the maximum backwater occurred as well as where the floodplain velocities were not retarded by the abutment. In general, this location varied between 67 and 133 percent of the bridge opening width measured upstream of the bridge. Depth-averaged velocities were determined at 17 to 19 positions across the cross section. These data were used to determine the discharge contraction ratio M at the beginning of scour as well as the approach velocity and depth upstream of the end of the abutment. Velocity and depth measurements were also made at the downstream face of the abutment, mostly for compound channel A. At this location, the depth of flow in the floodplain was very small; therefore, the velocities were measured in the main channel only and the discharge was obtained from integration. The discharge in the floodplain was obtained from continuity as the difference between the measured total discharge and the main-channel discharge.

Resultant velocities were measured at the upstream face of the bridge for both compound channels A and B for the fixed-bed case. The resultant velocity direction was determined either by flow visualization or by rotating the velocity meter until a maximum reading was obtained. In the case of compound channel B, additional resultant velocities were measured in the bridge opening to find the maximum velocity near the face of the abutment.

For the long abutments (L_a/B_f = 0.88, 0.97, 1.0), an acoustic Doppler velocimeter (ADV) was used to measure three-dimensional velocity components in the main channel at stations beginning at the approach section and extending to the contracted section. The measuring volume for this instrument is a cylinder 9 mm high by 6 mm in diameter located a distance of 5 cm below the probe. The ADV transmits two sound pulses separated by a time lag and receives the reflected signals from particles in the flow at the location of the sampling volume. The instrument measures the Doppler shift between the frequencies of the transmitted and received signals based on the change in phase of the return signal from each pulse, which can be correlated with velocity in the flow. Some limitations associated with this probe include: (1) a minimum water depth of approximately 7 cm is required for submergence of the probe with allowance for the 5-cm distance to the sampling volume; (2) when operating in highly turbulent environments, the signal-to-noise ratio (SNR) may drop below accepted levels; (3) the probe cannot sample a location too close to a boundary because of the size of the sampling volume and possible interference with the boundary; and (4) the maximum sampling frequency is 25 hertz (Hz). The first limitation means that velocities near the free surface cannot be measured and that the probe is not useful for very shallow flows over the floodplain in this investigation. Thus, the ADV was used only in the main channel where depths were on the order of 20 to 25 cm. The center of the sampling volume was limited to a minimum distance of 5 mm from the channel bottom. The SNR varied from 5 to 15 decibels (dB), which is adequate for measuring mean current speed; however, for SNR < 15, the manufacturer does not recommend high-resolution measurements (i.e., sampling at 25 Hz, which is required for turbulence measurements).

SCOUR MEASUREMENTS

After the fixed-bed measurements were completed, a movable-bed section was constructed in the vicinity of the abutments. The cement-stabilized floodplain surface was removed between stations 8.5 and 11.0 for compound channel A, and between stations 6.7 and 12.8 in the case of compound channel B. This resulted in a total length of movable bed for compound channel A of 2.5 m and a length of 6.1 m for compound channel B. The movable-bed floodplain consisted of sediments A, B, or C, with sizes as indicated previously in table 1. In both compound channels A and B for relative abutment lengths less than 0.66, the main channel remained as a fixed bed; for the longer abutments in compound channel B, the main channel was also made into a movable bed. It was formed using a template to produce the same slope as for the fixed bed. The abutments were sealed with cement mortar to the floor of the flume.

Scour measurements were made for several discharges at each of the abutment lengths given in table 1. At the start of each experimental scour test, the loose gravel in the movable-bed test section was carefully leveled to the floodplain elevations corresponding to the constant-bed slope. Water was then introduced into the channel very gradually from the upstream and downstream ends of the flume with the tailgate raised so that the movable bed remained undisturbed. Once the whole channel was flooded, the desired discharge was obtained by adjusting the valve on the main inlet pipe. The tailgate was then lowered slowly until the corresponding normal depth was reached at the downstream end of the channel. Scour was allowed to continue for 12 to 16 hours for compound channel A, and for 24 to 65 hours for compound channel B. After equilibrium had been reached, the water-surface profile and the velocity distributions in the approach section and at the downstream face of the abutment were measured in the case of compound channel B. Then the channel was carefully drained and the bed elevations throughout the scour and deposition areas were measured with a point gauge having a scale uncertainty of ±0.03 cm. As a practical matter, the uncertainty in the scour-depth measurements was about ±0.09 cm. In general, bed elevations in the vicinity of the scour hole were measured at approximately 100 spatial locations from which scour contours could be plotted.

The time rates of scour measurements were made for sediment A and for the vertical-wall abutment for some experiments. For these runs, scour depth was measured from a scale inscribed on a Plexiglas® block that formed the face of the abutment. The scale could be read from above by the use of a mirror. Only those cases where the maximum scour depth occurred at the upstream corner of the abutment face were suitable for this measurement technique. In general, these cases occurred for the greater discharges for each abutment length. Scour-depth measurements were generally taken at times of 0.08, 0.17, 0.25, 0.50, 0.75, 1, 2, 4, 8, 16, 24, and 36 hours after the beginning of the experiment.

RESULTS

Channel Roughness

The experimental results for Manning's n in the main channel and the floodplain with compound-channel flow are shown in figure 4 for sediment A and compound channel B. First,

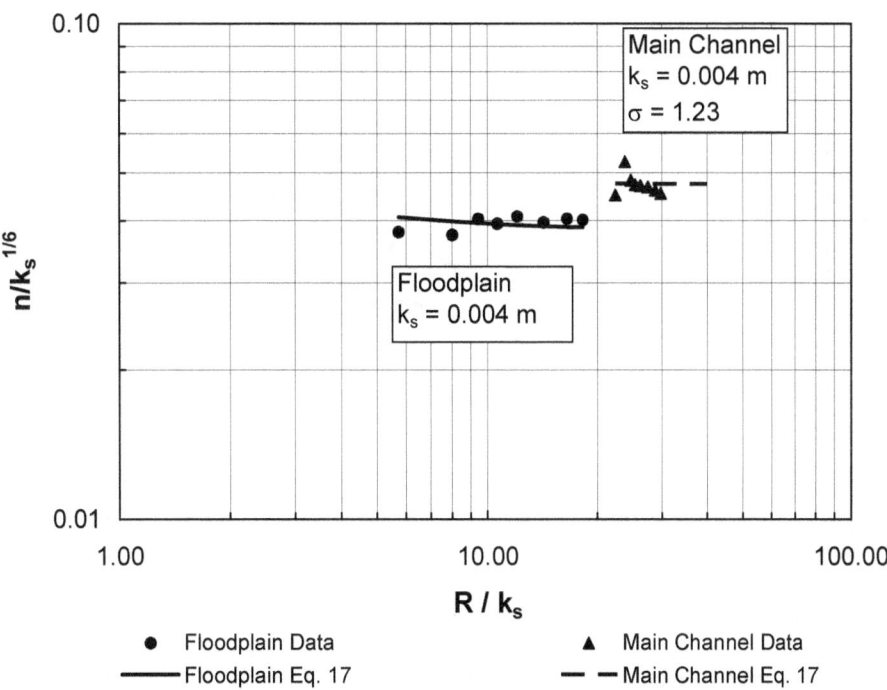

Figure 4. Manning's n in the main channel and floodplain for compound channel B.

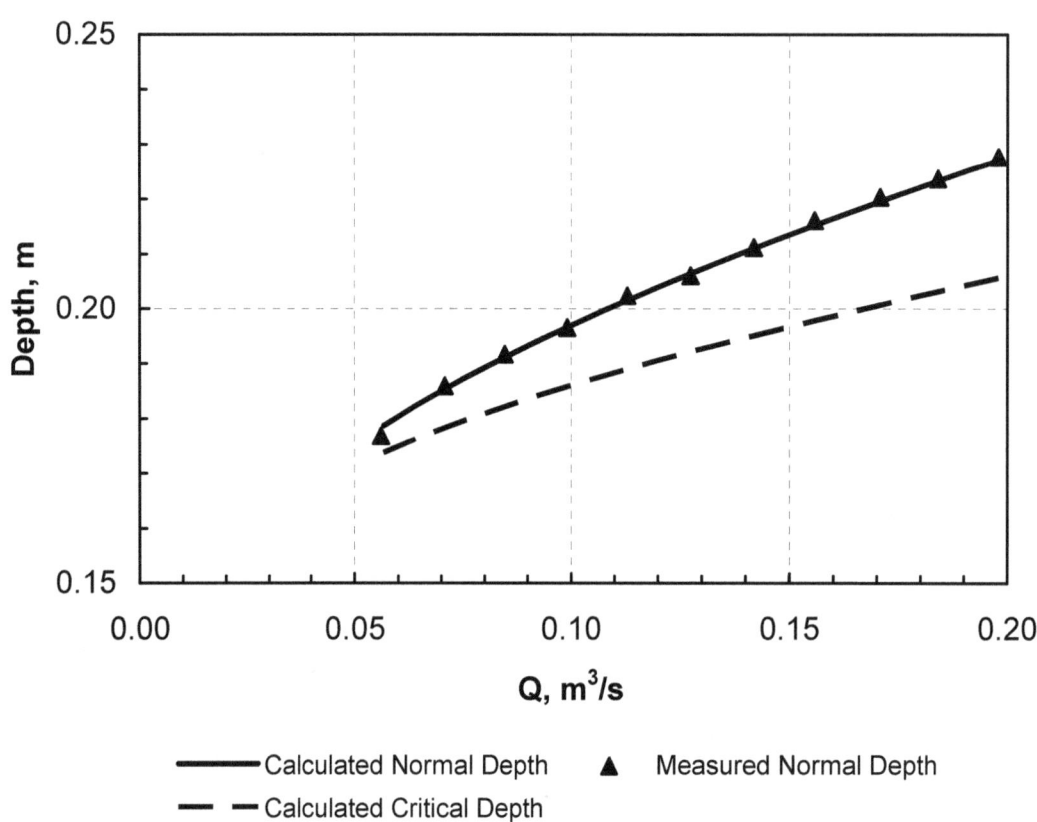

Figure 5. Measured and computed normal depth for compound channel B.

the value of the equivalent sand-grain roughness in the main channel was determined from the normal depth measurements for flow in the main channel alone (without overbank flow) with the result that $k_s = 0.004$ m (4.0 mm or $1.2d_{50}$). Then the separate discharges were determined in the main channel and the floodplain for overbank flow at several discharges as described previously, and the data points are shown in figure 4 in comparison with Keulegan's equation, given as:

$$\frac{n}{k_s^{1/6}} = \frac{0.113 \ (\frac{R}{k_s})^{1/6}}{2.03 \ \log \ (\frac{12.2 \ R}{k_s})} \tag{17}$$

where R = hydraulic radius in meters and k_s = equivalent sand-grain roughness in meters. For the separate flow in the floodplain, k_s was found to be 0.004 m as in the main-channel flow alone, which is consistent with the fact that the roughness surfaces were identical. In the case of main-channel flow, an additional drag caused by the main channel/floodplain interaction was quantified in terms of a coefficient σ, defined as:

$$n_{mc} = \sigma \ n_K \tag{18}$$

where n_{mc} = effective Manning's n in the main channel with overbank flow and n_K = Keulegan's value for Manning's n in the main channel with $k_s = 0.004$ m and no interaction with the floodplain (i.e., zero contribution of the interface to the wetted perimeter). The best-fit value of σ was 1.23. In effect, the main-channel conveyance was decreased by 23 percent because of the main channel/floodplain interaction. Similar results were found previously for compound channel A by Sadiq[22] and Sturm and Sadiq[40] with σ = 1.19. Shown in figure 5 are the measured values of normal depth in comparison with the calculated values for compound channel B using Keulegan's equation and the best-fit values of k_s and σ. The standard error in the normal depth is ±0.076 cm. While it is true that the values of Manning's n were first calibrated with the same data as in figure 5, the agreement shown in figure 5 is indicative of how well this calculation approach can reproduce the discharge Q over the full range of experimental values. Also shown in figure 5 is the calculated critical depth curve, which is below the normal depth curve, demonstrating that the channel is mild over the experimental range of discharges.

Discharge Distribution

The purpose of considering the main channel/floodplain interaction in developing Manning's n values for compound-channel flow is illustrated in figure 6 in which the ratio of the main-channel discharge to the total discharge is given as a function of relative depth in the floodplain. The data in figure 6 were measured in uniform flow without the presence of the abutment. In both compound channels A and B, the calculated discharge distribution agrees well with the measured values. For comparison, the standard WSPRO method of assuming a constant Manning's n and no interaction between the main channel and the floodplain overestimates the relative proportion of main-channel discharge as shown in figure 6 for compound channel A.

The correct proportioning of the main-channel and floodplain discharge is necessary to predict the discharge distribution factor M for which experimental values are given in figure 7(a) as a function of relative floodplain depth in the bridge approach section (indicated by a subscript of 1 on the depth).[1] The factor M is defined as:

$$M = \frac{Q_{m1} + (Q_{f1} - Q_{obst1})}{Q}$$

(19)

where Q_{m1} = discharge in the approach main channel, Q_{f1} = discharge in the approach floodplain, Q_{obst1} = obstructed floodplain discharge over a length equal to the abutment length in the approach cross section, and Q = total discharge, which is equal to $Q_{m1} + Q_{f1}$. While the values of M are primarily a function of the abutment length and the compound-channel geometry, figure 7(a) shows that they decrease slowly with increasing relative depth in the approach cross section. The abutment shape seems to have relatively little influence on M in figure 7(a).

Sturm and Janjua[19] have shown that for small depth changes from the approach section to the bridge section, M represents the ratio of discharge per unit width in the approach floodplain to that in the contracted floodplain in the bridge section, q_{f1}/q_{f2}. This discharge ratio has a significant effect on the equilibrium abutment scour depth as will be shown in chapter 4. Alternatively, it can be shown by the same reasoning that M is indicative of the value of the ratio of main-channel discharge in the approach section to that in the contracted section, Q_{m1}/Q_{m2}. The experimental results in figure 7(b) confirm that such a relationship exists. The value of M, and hence q_{f1}/q_{f2}, is not the same as the geometric contraction ratio m, which is defined as the bridge opening width over the total channel width. This is because of the characteristic behavior of compound-channel flow in which some of the approach floodplain flow joins the main-channel flow in the bridge section. The redistribution of flow between the main channel and the floodplain in the contracted bridge section has a significant influence on the scour depth as will be discussed further in chapter 4.

Raw experimental results for water-surface profiles, velocity distributions, critical velocity, and equilibrium scour depths, as well as scour contours for compound channel B, are summarized in the following sections of this chapter.

---

[1] In the remainder of this chapter and the rest of the report, the subscript "0" refers to depths, velocities, and discharges measured in uniform flow at the bridge location without the presence of an obstruction. The subscript "1" refers to the approach section upstream of the bridge with the obstruction in place where bridge backwater is at a maximum and where floodplain velocities are not retarded by the embankment. The subscript "2" represents the contracted flow section at the downstream face of the bridge. Main-channel variables are indicated by the subscript "m" and floodplain variables are denoted by the subscript "f." For example, y_{f1} is the floodplain depth in the approach section with the bridge in place. See figure 23.

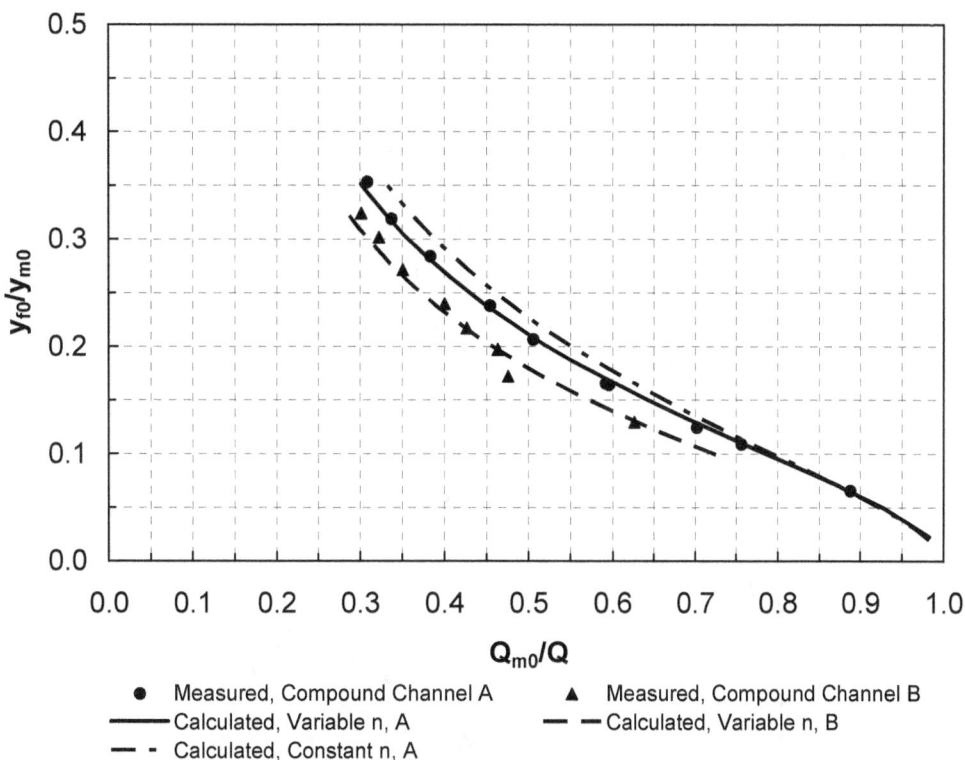

Figure 6. Ratio of main-channel discharge to total discharge as a function
of relative normal depth in the floodplain.

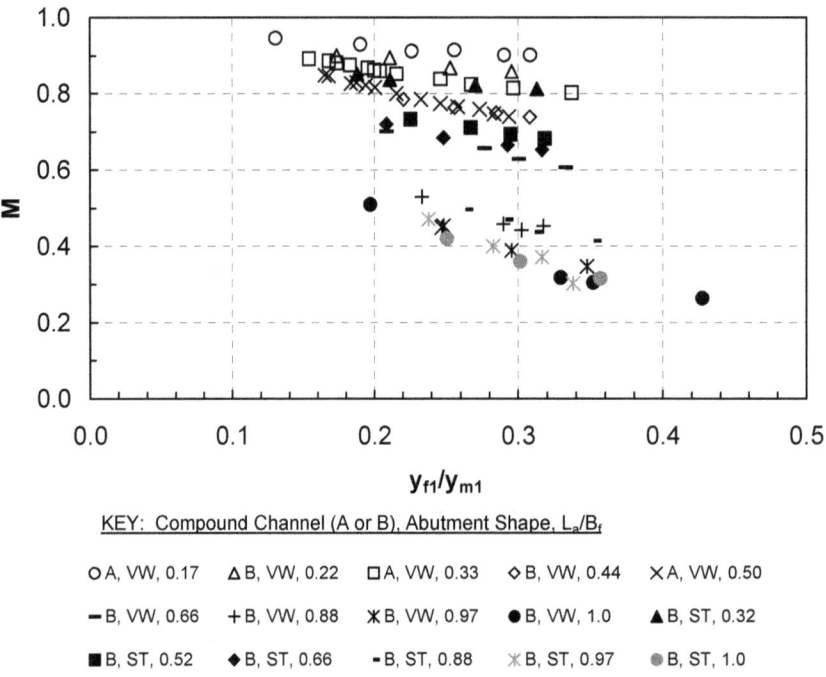

KEY: Compound Channel (A or B), Abutment Shape, L_a/B_f

○ A, VW, 0.17	△ B, VW, 0.22	□ A, VW, 0.33	◇ B, VW, 0.44	✕ A, VW, 0.50
− B, VW, 0.66	+ B, VW, 0.88	✳ B, VW, 0.97	● B, VW, 1.0	▲ B, ST, 0.32
■ B, ST, 0.52	◆ B, ST, 0.66	− B, ST, 0.88	✳ B, ST, 0.97	● B, ST, 1.0

(a) Dependence of M on approach depth and abutment length.

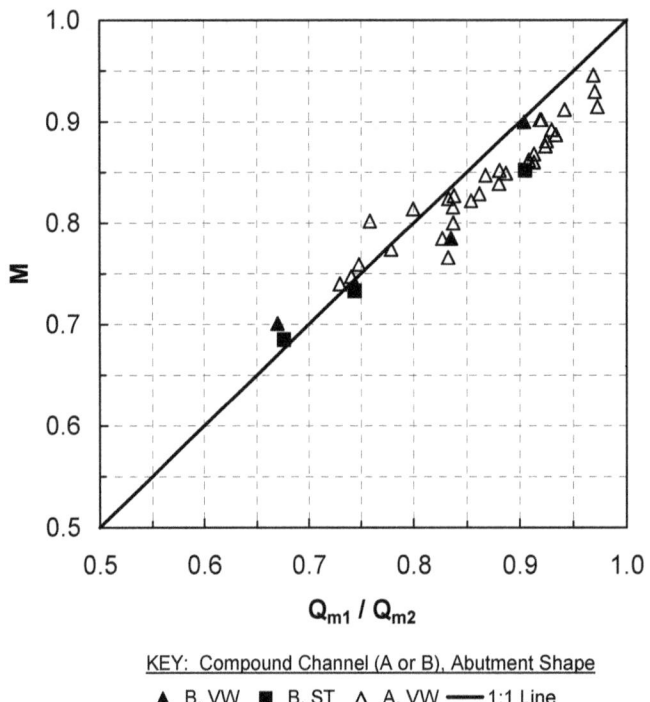

KEY: Compound Channel (A or B), Abutment Shape

▲ B, VW ■ B, ST △ A, VW —— 1:1 Line

(b) Use of M to reflect main-channel discharge ratio.

Figure 7. Dependence and use of discharge distribution factor M.

Water-Surface Profiles and Approach Velocity Distributions

Measured water-surface profiles and corresponding approach-section velocity distributions are shown in figures 8 and 9 for the vertical-wall abutment and the fixed-bed case (beginning of scour). In figure 8, the discharge is held constant at 0.0850 m^3/s and the relative abutment length is varied from 0.22 to 0.66. The downstream boundary condition is the same for each profile because it is the normal depth for the given discharge as set by the tailgate. The increasing backwater upstream of the abutment that is caused by increasing abutment length is apparent in figure 8(a). In figure 8(b), the main-channel velocities are higher than those in the floodplain as expected for small floodplain depths in a compound channel. The increasing backwater for longer abutments shown in figure 8(a) causes a general decrease in both the floodplain and main-channel velocities across the full width of the compound channel as shown in figure 8(b).

In figure 9, the relative abutment length is held constant at 0.44; however, the discharge is varied from 0.0708 m^3/s to 0.0991 m^3/s. In this case, the downstream boundary condition of normal depth causes an increasing tailwater depth with increasing discharge as can be seen in figure 9(a). The backwater upstream of the bridge abutment is also observed to increase with increasing discharge as seen in figure 9(a). In contrast to the effect of increasing abutment length at constant discharge, increasing discharge for the vertical-wall abutment of constant length causes an increase in the floodplain velocities as shown in figure 9(b), even though the backwater is increasing. The main-channel velocities also increase, but this occurs chiefly in the sideslope region of the main channel. The increase in the floodplain velocities with increasing discharge is a consequence of a greater proportion of the discharge being carried in the floodplain relative to the main channel as the relative depth increases. The change in flow distribution is also apparent in figure 7(a), in which M decreases with increasing relative floodplain depth because of the decrease in the unobstructed discharge for the same abutment length. Results similar to those of figures 8 and 9 were observed for the spill-through abutments. The relative degree of upstream backwater for vertical-wall and spill-through abutments is most easily summarized in terms of the ratio of the floodplain normal depth, y_{f0}, which would occur with no abutments, to the approach floodplain depth, y_{f1}. This backwater ratio varied from 0.6 to 1.0 as a function of abutment length and shape, and discharge.

For the larger abutment lengths, the scour depth increased even at the lowest discharges to the point that it was possible for the scour hole to bottom out on the concrete floor of the flume. To avoid this condition, the tailwater was raised above the normal depth so that the tailwater level became the independent variable for the scour depth while holding the discharge constant for a given abutment length. The effects of variable tailwater on the water-surface profiles and the approach velocity distributions are shown in figure 10. This is the case of the abutment terminating at the edge of the main channel ($L_a/B_f = 1.0$) for a constant discharge of 0.0567 m^3/s. It is clear from figure 10(a) that the higher the tailwater level, the less backwater upstream of the abutment, where backwater is defined relative to an upstream depth that would occur without the contraction. Beginning at the tailgate, the floodplain depth decreases in the upstream direction as an M1 profile because of boundary resistance both downstream and upstream of the abutment; however, the largest drop in depth occurs through the contracted section. The higher tailwater also causes less difference between the floodplain and main-channel velocities with the velocity distribution becoming nearly uniform for the highest tailwater as shown in figure 10(b). The

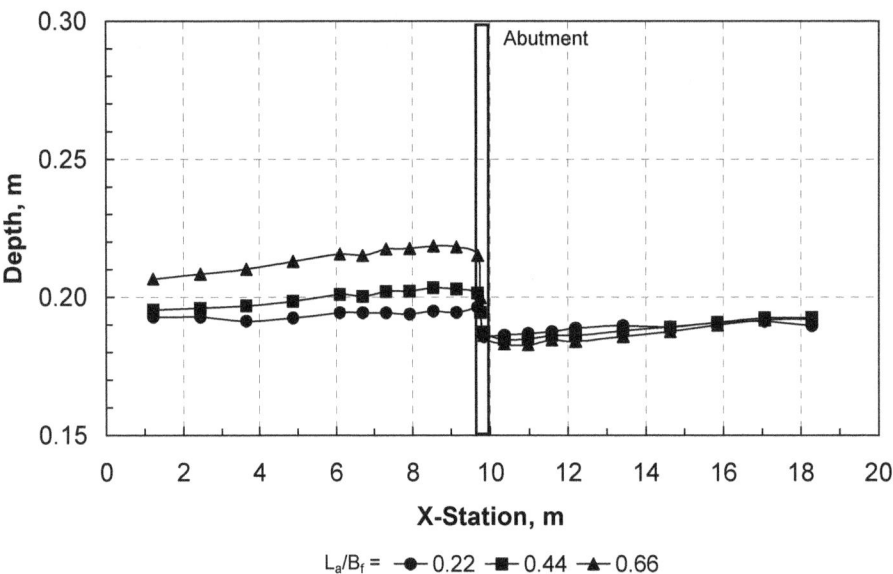

(a) Compound Channel B: Water-Surface Profiles,
VW Abutment, Q = 0.0850 m³/s.

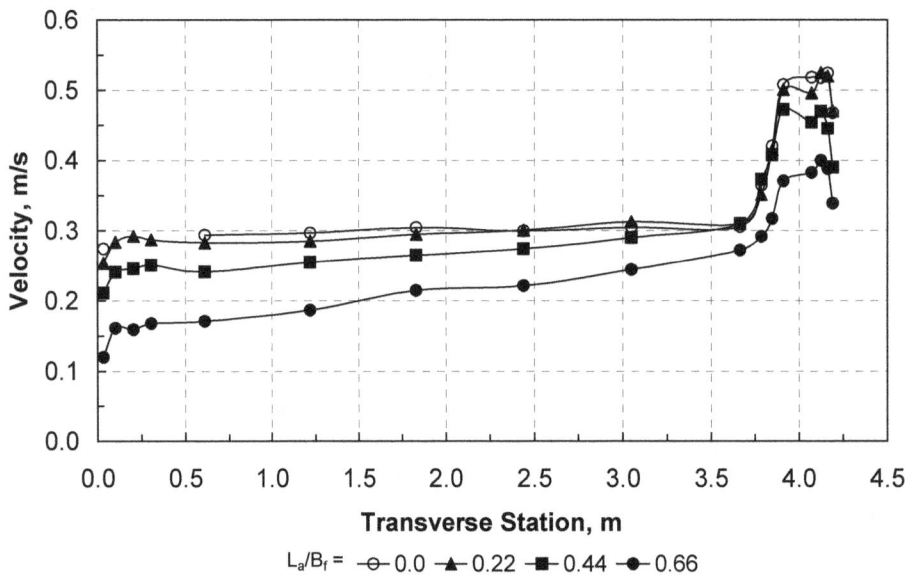

(b) Compound Channel B: Approach Velocity Distributions,
Station 7.3 m, VW Abutment, Q = 0.0850 m³/s.

Figure 8. Effect of variable abutment length on water-surface profiles
and velocity distributions for constant discharge.

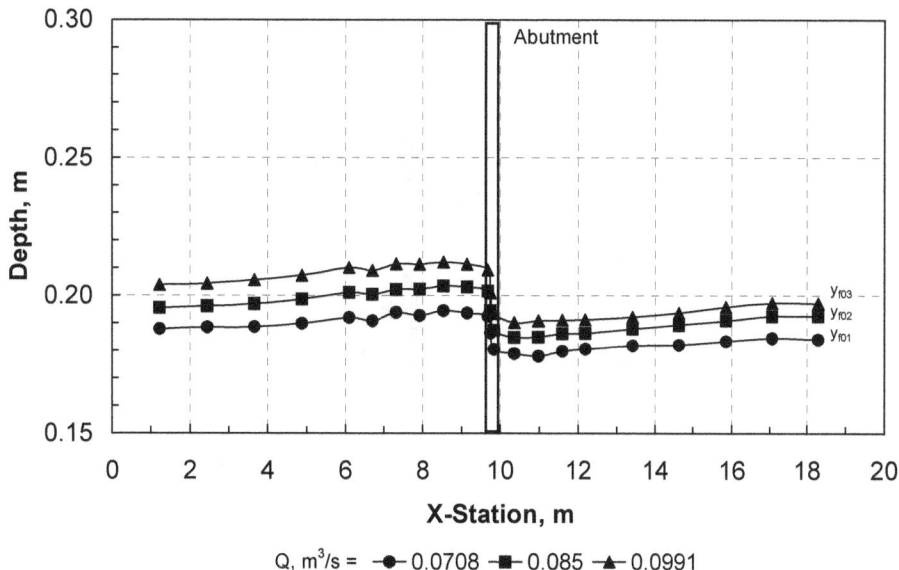

(a) Compound Channel B: Water-Surface Profiles,
VW Abutment, $L_a/B_f = 0.44$.

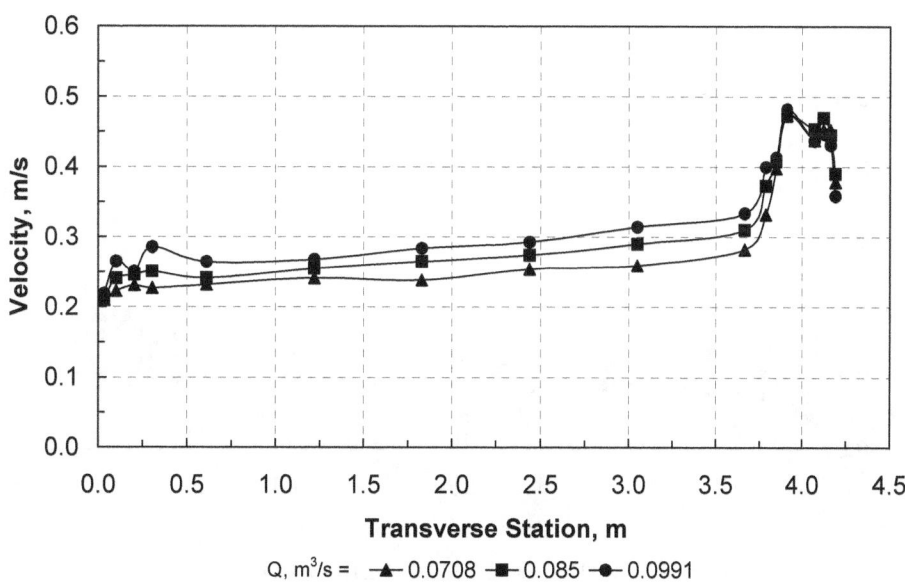

(b) Compound Channel B: Approach Velocity Distributions,
Station 7.3 m, VW Abutment, $L_a/B_f = 0.44$.

Figure 9. Effect of variable discharge on water-surface profiles
and velocity distributions for constant abutment length.

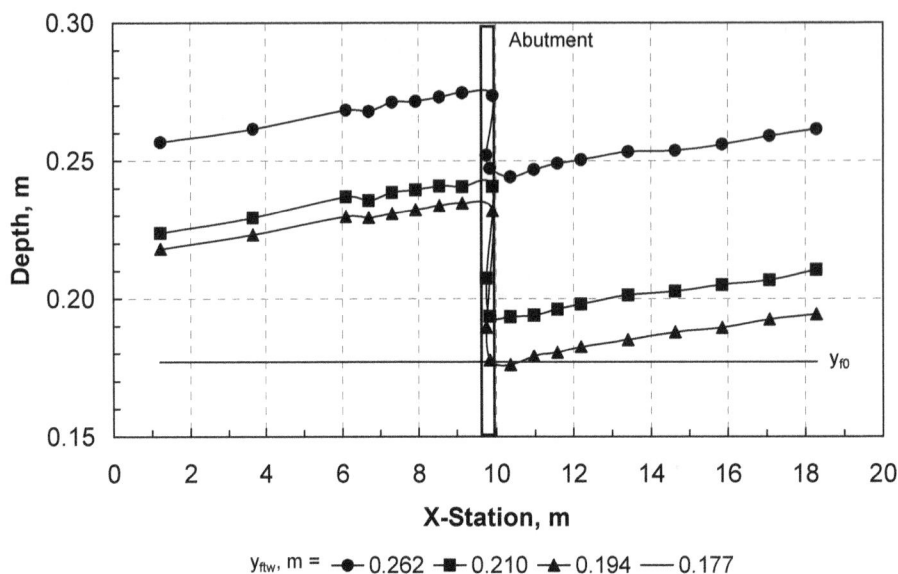

(a) Compound Channel B: Water-Surface Profiles,
VW Abutment, L_a/B_f = 1.0, Q = 0.0567 m^3/s.

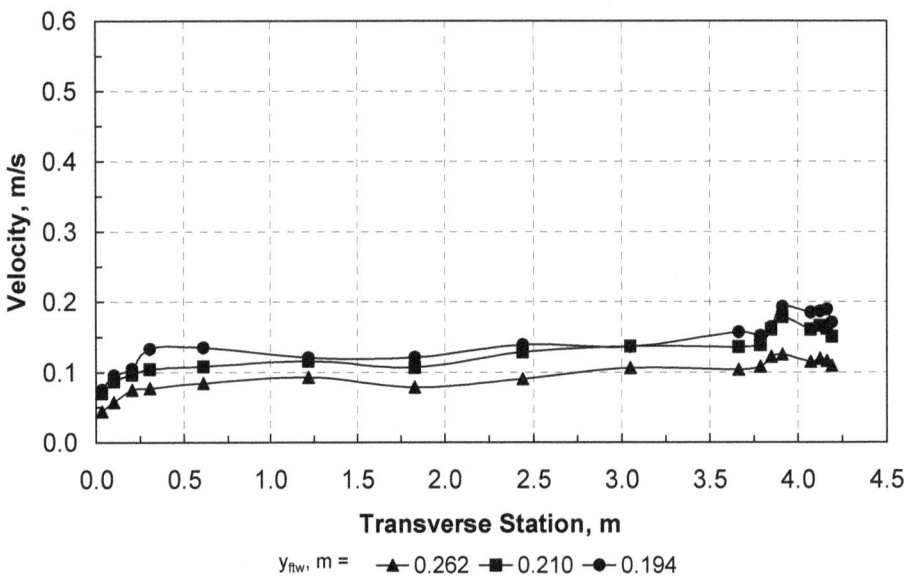

(b) Compound Channel B: Approach Velocity Distributions,
VW Abutment, L_a/B_f = 1.0, Q = 0.0567 m^3/s.

Figure 10. Effect of tailwater (y_{ftw}) on water-surface profiles and
velocity distributions for constant discharge.

highest tailwater in figure 10 corresponds to a relative floodplain depth of $y_f/y_m = 0.27$, at which the floodplain/main-channel velocity interactions become minimal, but the discharge per unit width must still be significantly larger in the main channel than the floodplain because of the larger depths there. Hence, even at these larger depths, the compound channel in these experiments cannot be treated as a rectangular channel in which the discharge per unit width is uniform across the channel.

As scour occurs over time, the hydraulic conditions at the contracted section change to the extent that the scour hole is large enough to alter the channel geometry. Water-surface profiles were measured before and after scour and some representative results are compared in figures 11 and 12. The water-surface profiles before scour were measured for a fixed bed of the same sediment as the movable bed, while the water-surface profiles after scour were measured after the scour hole had reached equilibrium. In figure 11, water-surface profiles before and after scour are compared for relative abutment lengths of 0.44 and 0.88. In figure 11(a), the tailwater is equal to the normal depth, while for the longer abutment shown in figure 11(b), the discharge is less and the tailwater is higher than the normal depth. The scour depths in figures 11(a) and 11(b) are 13.4 cm and 20.1 cm, respectively. Thus, the combination of a larger scour depth and wider scour hole in a narrower contracted section causes a marked decrease in the water depths upstream of the abutment after scour as shown in figure 11(b), while the change in the water-surface profile shown in figure 11(a) is much smaller. A similar comparison is shown in figure 12 for abutments that encroach on the main channel ($L_a/B_f = 0.97$ and 1.0). In comparison with figure 11(b), the water-surface profile drops much more after scour as shown in figure 12(a), even though the water discharge and the scour depths are the same (20.1 cm). This is probably caused by the increased backwater and narrower contraction as shown in figure 12(a). Upon comparison of figures 12(a) and 12(b), the scour depth in figure 12(b) is 18.3 cm, which is just slightly less than that in figure 12(a). The tailwater has been raised in figure 12(b) and the abutment is slightly longer. Apparently, these differences are sufficient to cause a greater water-surface drop after scour, almost to the level of the tailwater elevation downstream as shown in figure 12(b). All of the scour and all of the flow are forced to occur in the main channel of the contracted section for the conditions of figure 12(b).

Main-Channel Centerline Velocity Profiles

The ADV was used to measure centerline velocity profiles in the main channel starting from the approach section and continuing downstream to the contracted section. Velocities were measured in three dimensions, with the x-coordinate positive in the longitudinal flow direction, the y-coordinate positive upward, and the z-coordinate positive to the right when looking downstream. The x and z components of the velocity, V_x and V_z, are shown in figures 13, 14, and 15 for relative abutment lengths of 0.88, 0.97, and 1.0 as a function of the relative depth y'/y_m, where y_m = local depth in the main channel and y' = vertical coordinate taken positive upward from the bottom of the main channel. The vertical components of the velocity were positive downward as drawdown occurred toward the contracted section; however, they were an order of magnitude smaller than the z components (lateral components) of the velocity and are not shown in the figures. The stations (STA.) shown in figures 13, 14, and 15 refer to the distance in meters downstream from the channel entrance. The centerline of the abutment is located at STA. 9.8.

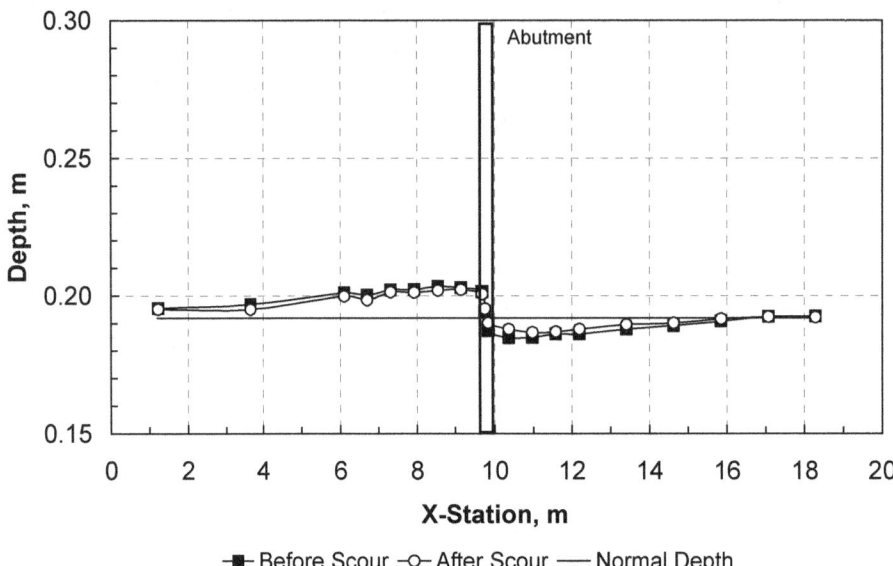

(a) Tailwater equal to normal depth,
Compound Channel B: Water-Surface Profiles,
VW Abutment, L_a/B_f = 0.44, Q = 0.0850 m³/s.

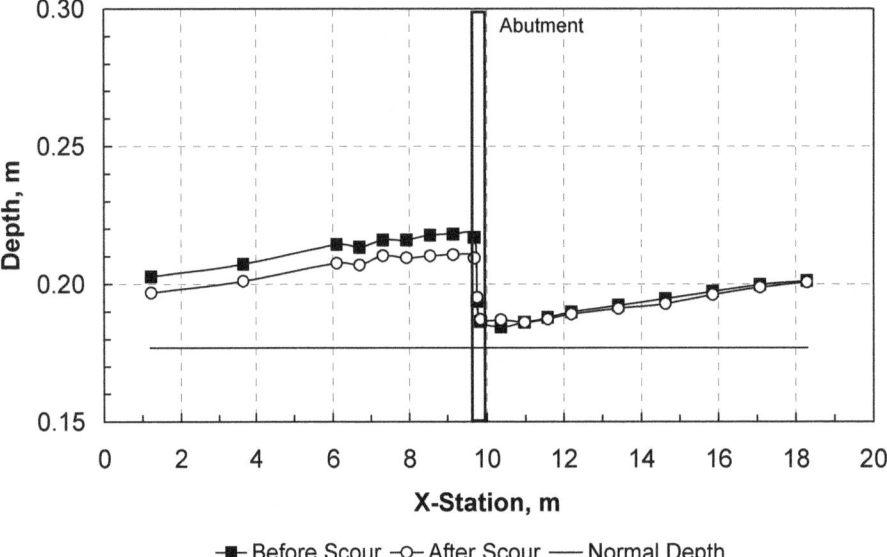

(b) Tailwater greater than normal depth,
Compound Channel B: Water-Surface Profiles,
VW Abutment, L_a/B_f = 0.88, Q = 0.0567 m³/s (B).

Figure 11. Water-surface profiles before and after scour for L_a/B_f = (a) 0.44 and (b) 0.88.

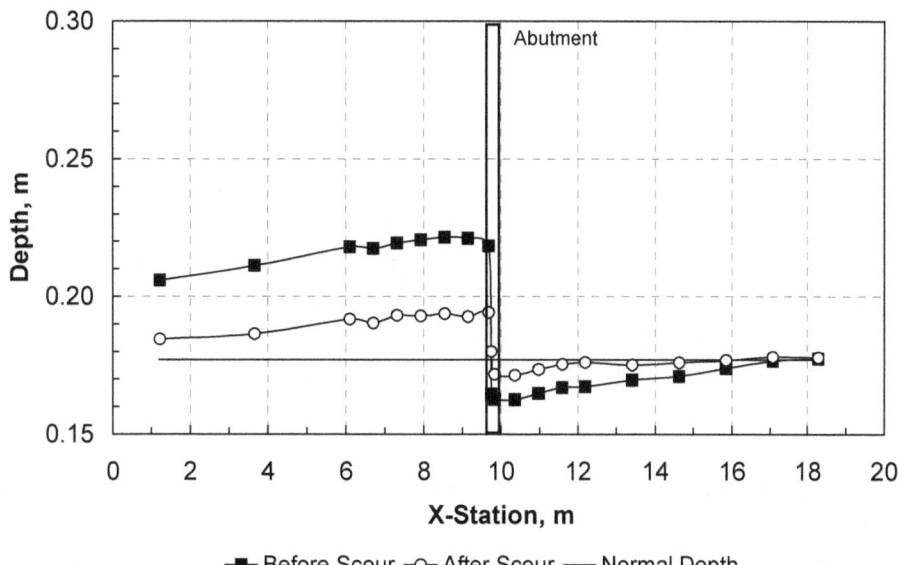

(a) Tailwater equal to normal depth,
Compound Channel B: Water-Surface Profiles,
VW Abutment, $L_a/B_f = 0.97$, $Q = 0.0567$ m^3/s.

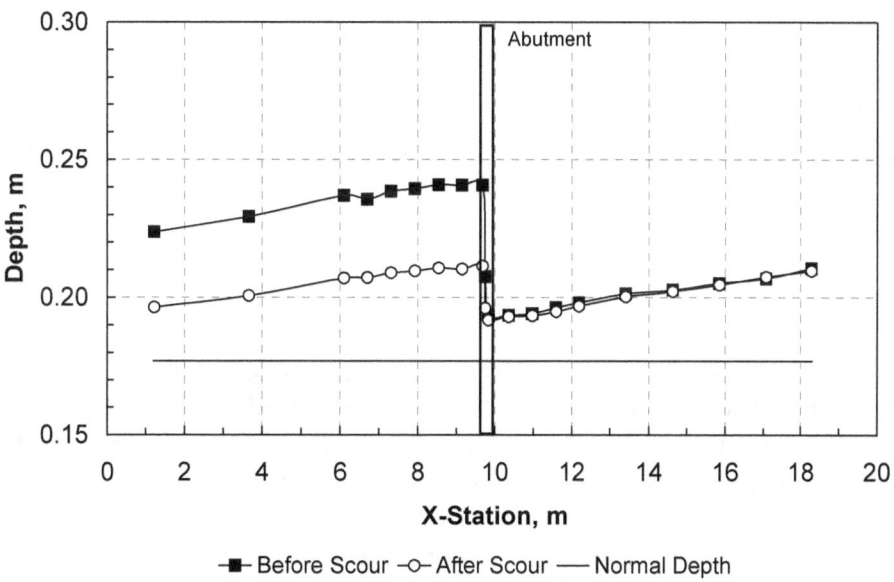

(b) Tailwater greater than normal depth,
Compound Channel B: Water-Surface Profiles,
VW Abutment, $L_a/B_f = 1.0$, $Q = 0.0567$ m^3/s (B).

Figure 12. Water-surface profiles before and after scour for L_a/B_f = (a) 0.97 and (b) 1.0.

The longitudinal velocities, V_x, in figures 13(a), 14(a), and 15(a), show that while there is significant acceleration in the main channel from STA. 6.1 to STA. 8.5, most of the acceleration occurs between STA. 8.5 and STA. 9.8. The flow from the floodplain joins the flow in the main channel as the contracted section is approached. This lateral flow is apparent in figures 13(b), 14(b), and 15(b), where its velocity, V_z, is observed to continuously increase as the contracted section is approached. The lateral velocities vary from 10 to 15 percent of the longitudinal velocities. The corresponding water-surface profiles have already been shown as figures 11(b), 12(a), and 12(b), and they show a very gradual increase in depth from STA. 6.1 to STA. 8.5 followed by a leveling off up to the upstream side of the embankment and then an abrupt drop into the contracted section itself. The three-dimensional velocities shown in figures 13, 14, and 15 make it clear that the lateral flux of flow from the floodplain into the main channel creates a driving mechanism for scour in addition to the horseshoe vortexes shed at the upstream edge of the abutment.

The magnitudes of the velocities in figures 13, 14, and 15 relative to the critical velocity for the initiation of motion suggest a dilemma for conducting live-bed scour experiments with long abutments in which sediment transport occurs only in the main channel upstream of the abutment. If the critical velocity of the sediment is approximately 50 cm/s, for example, then no sediment transport can occur in the approach section where the velocities in the main channel are approximately 20 cm/s; however, large scour depths can occur in the contracted section where the initial velocities are higher than 50 cm/s. An increase in discharge will increase the velocity in the contracted section and cause more scour; however, it will result in relatively small increases in the approach velocity in the main channel as shown previously in figure 9(b). On the other hand, increases in tailwater to limit the scour depth at the same discharge only serve to increase the critical velocity while decreasing the approach flow velocities. This leaves the alternative of a drastic reduction in sediment size to achieve a critical velocity of less than 20 cm/s, but then the scour depth will become significantly larger if maximum clear-water scour has not been reached, with the danger of bottoming out on the flume floor. Thus, the compound cross section, sediment size, and sediment thickness must be specifically designed for the case of live-bed scour.

Critical Velocity

Critical velocity for the initiation of sediment motion at the abutment face was measured for all three sediments. The tailwater was raised higher than the normal depth for a given flow rate, the flow rate was set, and then the tailgate was gradually lowered until the normal depth for that flow rate was reached. If sediment motion did not occur, then the tailgate was raised again and the flow rate was increased. This process was repeated until sediment motion had just begun at the face of the abutment. The determination of the conditions for the initiation of sediment motion necessarily involved some qualitative judgment; hence, there is scatter in the data points shown in figure 16. Only for sediment C was it possible to initiate motion on the floodplain with no abutments in place and for uniform flow. This experimental point is also shown in figure 16. Shown for comparison in figure 16 are the various relationships that can be used to calculate the critical conditions in terms of the critical value of the sediment number N_{sc}:

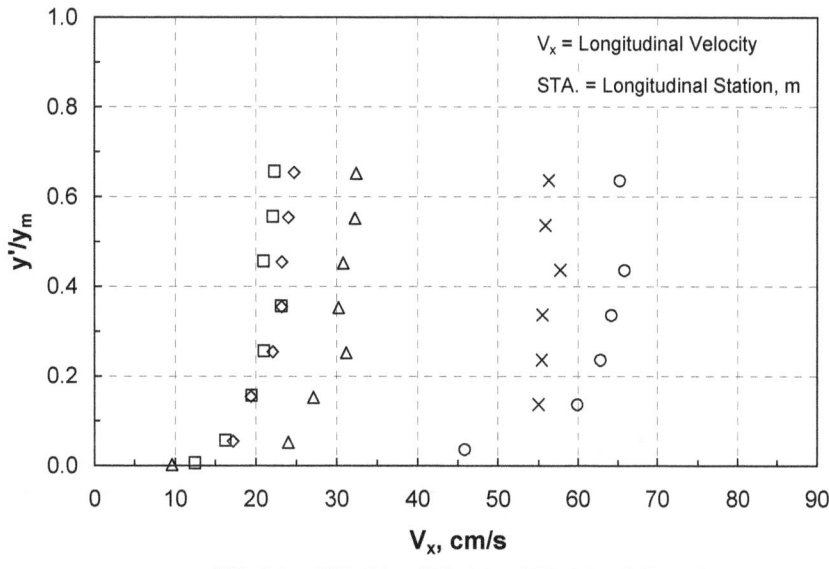

(a) Main-Channel Centerline Velocity-x,
VW Abutment, $L_a/B_f = 0.88$, $Q = 0.0567$ m^3/s (B).

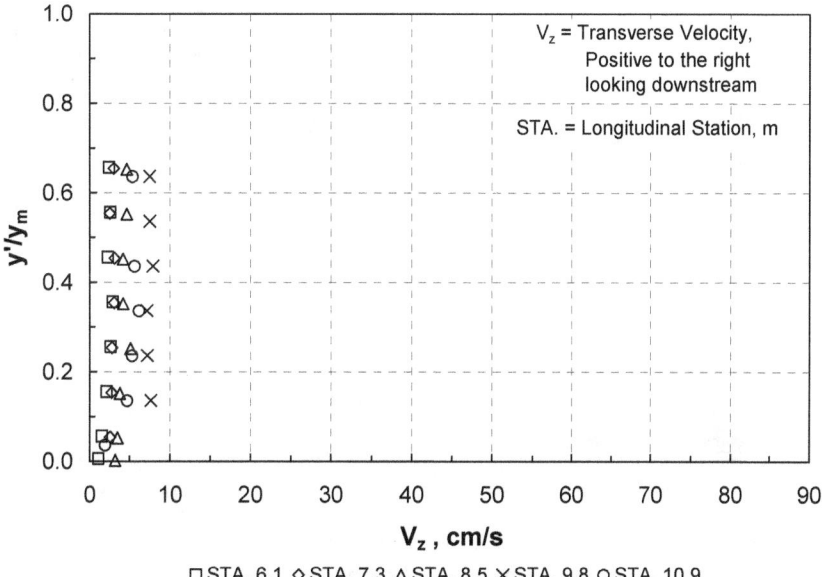

(b) Main-Channel Centerline Velocity-z,
VW Abutment, $L_a/B_f = 0.88$, $Q = 0.0567$ m^3/s (B).

Figure 13. Main-channel centerline velocity from approach
to contracted section for $L_a/B_f = 0.88$.

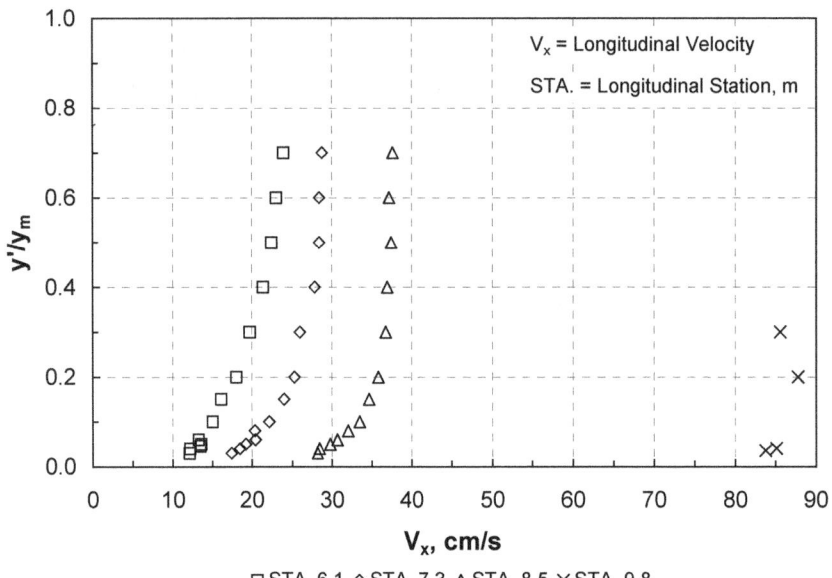

(a) Main-Channel Centerline Velocity-x,
VW Abutment, L_a/B_f = 0.97, Q = 0.0567 m^3/s.

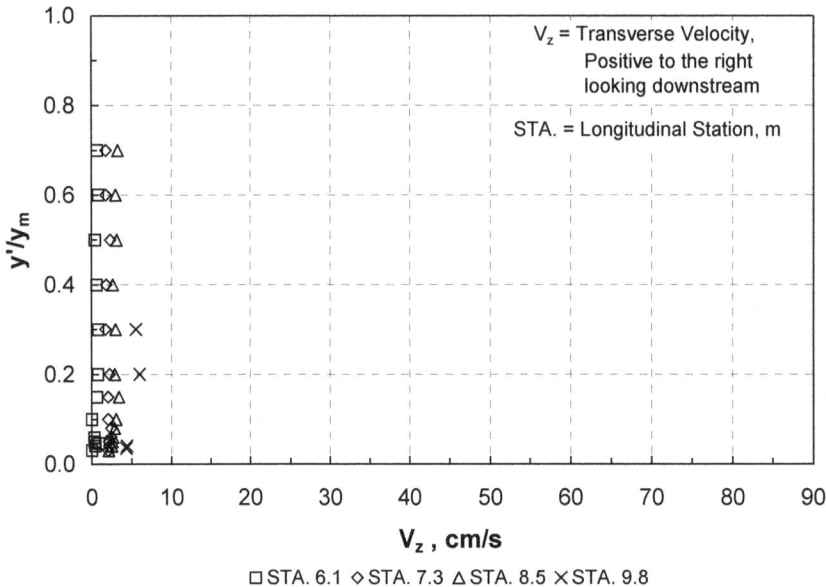

(b) Main-Channel Centerline Velocity-z,
VW Abutment, L_a/B_f = 0.97, Q = 0.0567 m^3/s.

Figure 14. Main-channel centerline velocity from approach
to contracted section for L_a/B_f = 0.97.

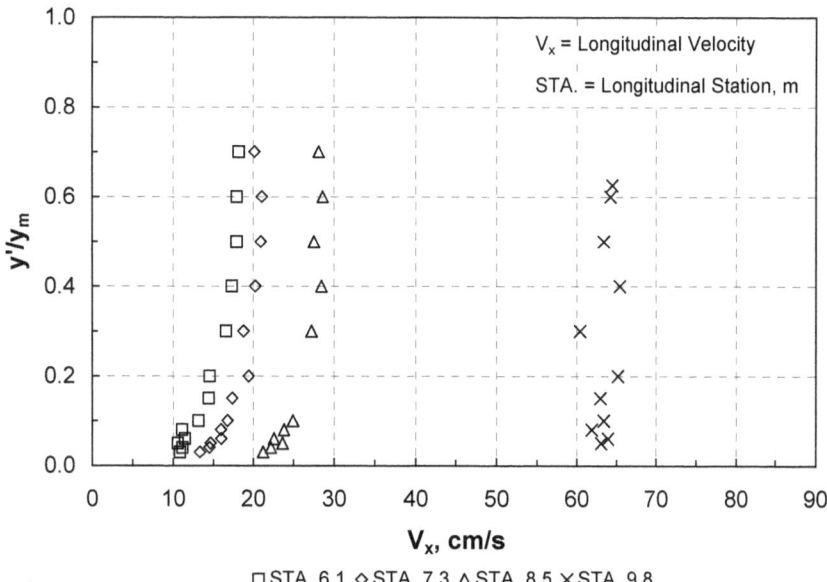

(a) Main-Channel Centerline Velocity-x,
VW Abutment, $L_a/B_f = 1.0$, $Q = 0.0567$ m³/s (B).

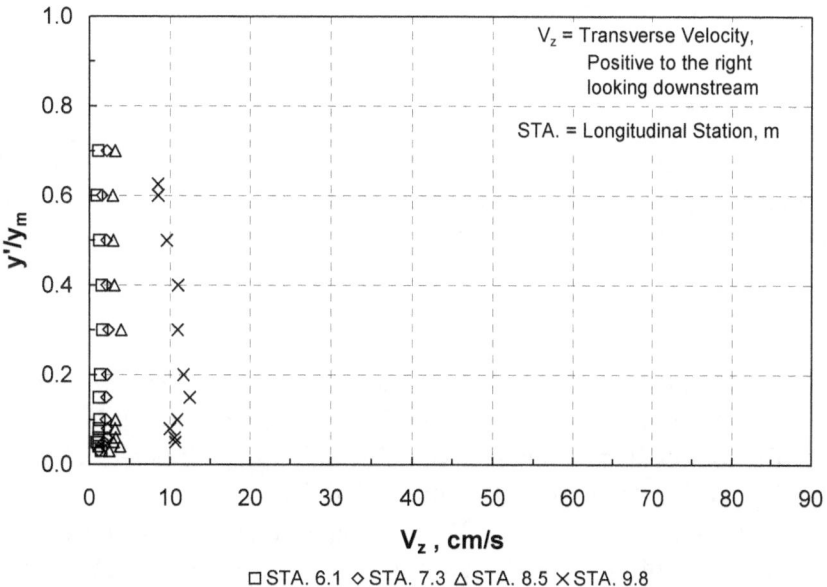

(b) Main-Channel Centerline Velocity-z,
VW Abutment, $L_a/B_f = 1.0$, $Q = 0.0567$ m³/s (B).

Figure 15. Main-channel centerline velocity from approach
to contracted section for $L_a/B_f = 1.0$.

$$PAROLA: \quad N_{sc} = \frac{1.61}{[d_{50} / y]^{0.135}} \tag{20}$$

$$LAURSEN: \quad N_{sc} = \frac{7.70 \sqrt{\tau_{*c}}}{[d_{50} / y]^{1/6}} \tag{21}$$

$$KEULEGAN: \quad N_{sc} = 5.75[\tau_{*c}]^{1/2} \log[\frac{12.2y}{2d_{50}}] \tag{22}$$

where $N_{sc} = V_c/[(SG - 1)gd_{50}]^{1/2}$, V_c = critical velocity, SG = specific gravity of sediment, d_{50} = median sediment grain diameter, y = flow depth, and τ_{*c} = critical value of Shields' parameter. Based on the results in figure 16 and the best fit of the scour data, which will be discussed in the next chapter, the uniform-flow critical velocity V_{c0} was calculated from Keulegan's relationship, with τ_{*c} determined from Shields' diagram for the given sediment size.[75] However, Laursen's relationship was found to give a better fit to the data for critical velocity at the abutment face (V_c), with τ_{*c} varying from 0.035 for sediment C to 0.039 for sediment A. (The value of 0.039 corresponds to $\tau_c = 4d_{50}$ in U.S. units.)

Equations 20, 21, and 22 can be used in either SI or U.S. units. Equation 20 is an empirical fit of the data taken by Parola on rock riprap in a laboratory channel as reported by Pagan-Ortiz,[85] while equation 22 follows directly from Keulegan's equation, with $k_s = 2d_{50}$ and the definition of the critical shear velocity in terms of Shields' parameter. However, the assumptions embedded in equation 21 can be clarified by outlining its derivation. As shown by Sturm and Sadiq,[23] expressing shear stress in terms of slope from the assumption of uniform flow and substituting for the slope from Manning's equation results in:

where $C_n = 1.486$ in U.S. units and 1.0 in SI units; k_n = constant in Strickler-type relationship for Manning's n ($n = k_n d_{50}^{1/6}$), which is equal to 0.0340 in U.S. units and 0.0414 in SI units; SG = specific gravity of the sediment; τ_{*c} = critical value of Shields' parameter; d_{50} = median grain diameter; and y = depth of uniform flow. Equation 21 follows directly from equation 23. The coefficient of 7.70 ($= C_n/k_n g^{0.5}$) in equation 21 is the same regardless of the system of units; however, it assumes the use of the Strickler constant. If some other constant is used, then the coefficient in equation 21 would change accordingly.

$$V_c = \frac{C_n}{k_n} \sqrt{(SG - 1) \tau_{*c}} \; d_{50}^{1/3} \; y^{1/6} \tag{23}$$

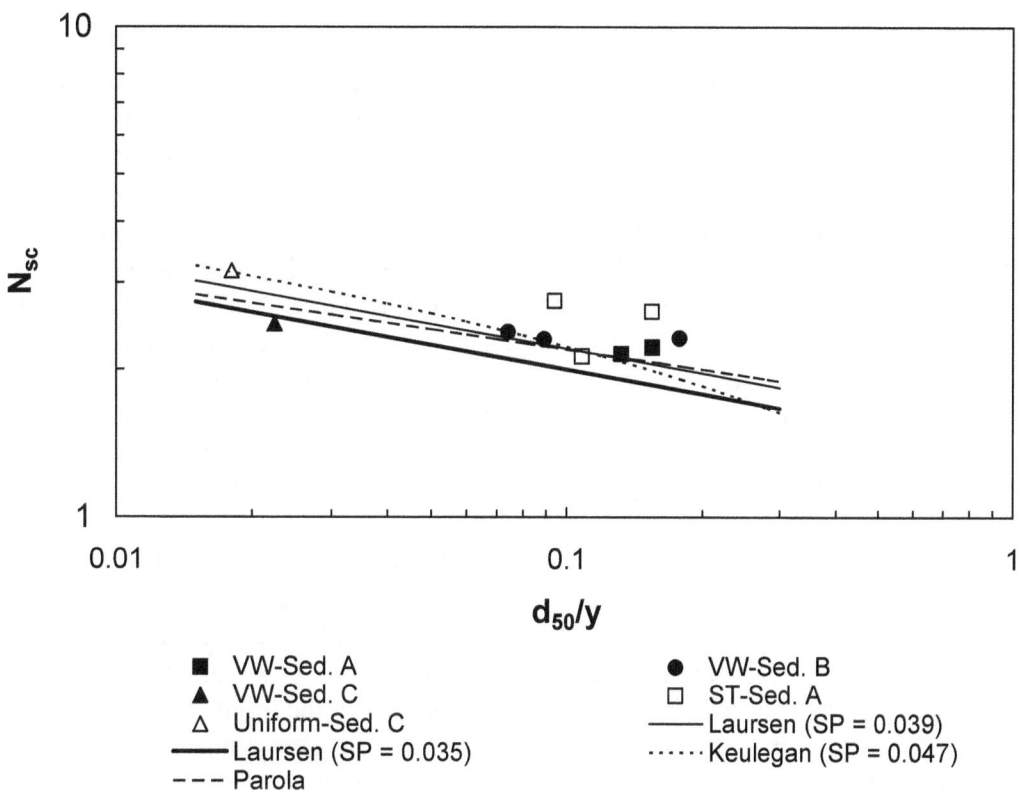

Figure 16. Measured and calculated critical velocities at incipient motion.
(SP = Shields Parameter).

Equilibrium Scour Depths

Raw data for the full set of equilibrium scour experiments are given in table 2. Several depths and velocities were measured prior to the beginning of scour for the fixed-bed case, and they are summarized in the table. The normal depth in the floodplain is y_{f0} and it varies with the discharge according to the uniform-flow measurements described previously in this chapter. In some instances, for the longer abutments, the tailgate was raised to submerge normal depth and limit the maximum scour. Under these conditions, the depth in the floodplain just upstream of the tailgate was measured and is reported as y_{ftw} to indicate a tailwater different from the normal depth. The approach velocity and depth, V_{f1} and y_{f1}, were measured in the floodplain upstream of the end of the abutment at the approach cross section, which was taken either as station 6.7 or 7.3 for $L_a/B_f \leq 0.66$, or as station 6.1 for $L_a/B_f \geq 0.88$, with the centerline of the abutment located at station 9.8. These distances were chosen so that the backwater was near the maximum value and the velocity distribution was unaffected by the contraction. The approach velocity in the floodplain was measured at a relative depth of 0.4 as an estimate of the depth-averaged velocity. The *resultant* velocity and depth near the upstream corner of the abutment face were measured and are given in table 2 as V_{ab} and y_{ab}, respectively. The measured approach velocity distribution was measured as described previously and integrated to produce values of the discharge contraction ratio M defined previously by equation 19. The measured velocity distributions were also integrated to determine the discharge in the main channel in ratio to the total discharge from the integration, and this ratio is given as Q_{m1}/Q in column 6 of table 2. The critical velocity, V_{0c}, was calculated from Keulegan's equation using the floodplain normal depth y_{f0} for relative abutment lengths, L_a/B_f, less than or equal to 0.88. For all longer abutments that encroached on the main channel, the depth used to obtain the critical velocity was the normal or tailwater depth in the main channel. The scour depths, d_s, shown in table 2 were all measured downward from the floodplain elevation at the abutment. They vary from 1.2 cm to 31.7 cm, and show a consistent increase with discharge. The smaller the sediment size, the larger the scour depth, as expected for clear-water scour. It should be noted that these scour experiments cover a range of values of L_a/y_{f1} from 12 to 97, and thus cover both intermediate-length and long abutments according to Melville's classification. The last column in table 2 shows the scour test duration, which varied from 9.5 to 65.6 hours with a mean value of 34 hours. In all but one case, the duration equaled or exceeded 24 hours.

Typical scour contours are shown in figures 17 through 22. The contours are given in terms of bed elevation in centimeters. The undisturbed floodplain and main-channel elevations at the station corresponding to the centerline of the abutment (station 9.75) are approximately 32.3 cm and 16.9 cm, respectively. A typical scour hole is shown in figure 17(a) for an intermediate vertical-wall abutment length ($L_a/B_f = 0.44$) and a discharge of 0.0992 m³/s. The bottom of the scour hole is displaced laterally from the abutment face and the long axis of the scour hole is skewed from the approach flow direction as an indication of the deflected streamlines in the contracted section. A prominent bar deposit can be seen on the downstream side of the abutment in the region where the flow is expanding. An increase in discharge to 0.117 m³/s is shown in figure 17(b), and the shape and location of the scour hole have changed. The maximum depth of scour occurs at the upstream corner of the abutment face, and the scour-hole shape is more

Table 2. Raw experimental results for fixed-bed hydraulic variables and equilibrium scour depth.

Shape	L_a, m	Q, m^3/s	d_{50}, mm	M	Q_{m1}/Q	y_{f0}, cm	y_{ftw}, cm	y_{f1}, cm	V_{f1}, cm/s	y_{ab}, cm	V_{ab}, cm/s	d_s, cm	time, hrs
VW	0.80	0.0700	3.3	0.900	0.51	3.29	3.29	3.23	25.63	2.65	49.7	1.2	9.5
VW	0.80	0.0850	3.3	0.895	0.45	3.87	3.87	4.11	29.35	3.14	57.5	3.7	29.3
VW	0.80	0.0994	3.3	0.883	0.40	4.42	4.42	4.60	30.75	3.66	60.3	11.0	51.0
VW	0.80	0.1133	3.3	0.868	0.37	4.94	4.94	5.21	33.16	4.05	64.9	16.5	32.5
VW	0.80	0.1278	3.3	0.865	0.35	5.39	5.39	5.88	34.62	4.72	68.4	19.5	29.8
VW	0.80	0.1425	3.3	0.859	0.33	5.85	5.85	6.46	35.84	5.36	71.1	22.9	34.0
VW	1.60	0.0637	3.3	0.790	0.51	2.97	2.97	3.96	23.71	2.53	61.3	6.1	30.4
VW	1.60	0.0714	3.3	0.785	0.47	3.29	3.29	4.36	24.81	2.74	63.5	10.1	31.4
VW	1.60	0.0705	3.3	0.785	0.47	3.29	3.29	4.36	24.81	2.74	63.5	8.8	37.7
VW	1.60	0.0850	3.3	0.764	0.41	3.87	3.87	5.30	26.21	3.05	71.0	13.4	35.5
VW	1.60	0.0992	3.3	0.749	0.37	4.42	4.42	6.13	28.22	3.60	77.9	19.8	32.2
VW	1.60	0.0992	3.3	0.749	0.37	4.42	4.42	6.13	28.22	3.60	77.9	17.7	55.4
VW	1.60	0.1028	3.3	0.744	0.36	4.69	4.69	6.49	28.86	3.69	78.1	17.7	29.4
VW	1.60	0.1130	3.3	0.739	0.34	4.94	4.94	6.86	29.59	3.81	78.6	22.9	35.0
VW	1.60	0.1173	3.3	0.729	0.33	5.18	5.18	7.35	30.27	4.02	83.0	28.3	28.8
VW	2.40	0.0564	3.3	0.701	0.50	2.63	2.63	4.05	19.38	1.22	69.9	8.5	50.1
VW	2.40	0.0637	3.3	0.679	0.46	2.97	2.97	4.85	20.15	1.55	76.2	15.5	33.2
VW	2.40	0.0708	3.3	0.657	0.43	3.29	3.29	5.88	20.79	1.92	83.1	18.3	35.0
VW	2.40	0.0850	3.3	0.629	0.38	3.87	3.87	6.61	22.65	2.59	91.3	23.8	36.4
VW	2.40	0.0856	3.3	0.629	0.38	3.87	3.87	6.61	22.65	2.59	91.3	20.7	35.7
VW	2.40	0.0994	3.3	0.607	0.34	4.42	4.42	7.68	24.99	2.90	98.5	29.0	33.1
VW	3.23	0.0499	3.3	0.530	0.46	2.26	2.32	4.69	16.82	2.26	69.5	20.1	33.0
VW	3.23	0.0567	3.3	0.452	0.36	2.50	6.46	7.16	15.39	5.49	76.8	18.0	34.0
VW	3.23	0.0567	3.3	0.458	0.37	2.50	4.88	6.28	16.55	3.87	69.6	21.6	27.5
VW	3.23	0.0567	3.3	0.442	0.35	2.50	5.52	6.67	16.37	4.42	74.2	25.3	27.0
VW	3.55	0.0496	3.3	0.453	0.42	2.26	2.26	5.09	15.82	1.22	83.0	17.1	27.2
VW	3.55	0.0565	3.3	0.389	0.36	2.50	2.50	6.46	16.95	1.46	94.7	20.1	24.0
VW	3.55	0.0671	3.3	0.347	0.31	3.05	3.32	8.20	15.88	1.65	108.3	29.3	24.0
VW	3.66	0.0569	3.3	0.264	0.26	2.50	10.91	11.49	10.39	10.67	63.0	15.5	28.5
VW	3.66	0.0569	3.3	0.305	0.31	2.50	5.79	8.35	13.59	6.71	77.3	18.3	26.8
VW	3.66	0.0567	3.3	0.318	0.32	2.50	4.21	7.56	15.70	5.49	83.9	27.4	27.0
VW	0.80	0.0856	2.7	0.895	0.45	3.87	3.87	4.11	29.35	3.14	57.5	10.7	31.8
VW	0.80	0.0992	2.7	0.883	0.40	4.42	4.42	4.60	30.75	3.66	60.3	14.3	55.5
VW	0.80	0.1144	2.7	0.868	0.37	4.94	4.94	5.21	33.16	4.05	64.9	19.5	32.1
VW	0.80	0.1278	2.7	0.865	0.35	5.39	5.39	5.88	34.62	4.72	68.4	21.3	26.7
VW	1.60	0.0711	2.7	0.785	0.47	3.29	3.29	4.36	24.81	2.74	63.5	12.2	65.6
VW	1.60	0.0853	2.7	0.764	0.41	3.87	3.87	5.30	26.21	3.05	71.0	14.6	33.0
VW	1.60	0.0992	2.7	0.749	0.37	4.42	4.42	6.13	28.22	3.60	77.9	20.1	33.8
VW	1.60	0.1065	2.7	0.744	0.35	4.69	4.69	6.49	28.86	3.69	78.1	27.1	35.3
VW	2.40	0.0567	2.7	0.701	0.50	2.63	2.63	4.05	19.38	1.22	69.9	12.5	35.5

Table 2. Raw experimental results for fixed-bed hydraulic variables and equilibrium scour depth (continued).

Shape	L_a, m	Q, m^3/s	d_{50}, mm	M	Q_{m1}/Q	y_{f0}, cm	y_{ftw}, cm	y_{f1}, cm	V_{f1}, cm/s	y_{ab}, cm	V_{ab}, cm/s	d_s, cm	time, hrs
VW	2.40	0.0637	2.7	0.679	0.462	2.97	2.97	4.85	20.1	1.55	76.2	18.6	35.2
VW	2.40	0.0711	2.7	0.657	0.426	3.29	3.29	5.88	20.8	1.92	83.1	21.3	33.0
VW	2.40	0.0853	2.7	0.629	0.381	3.87	3.87	6.61	22.6	2.59	91.3	27.4	33.8
VW	0.80	0.0567	1.1	0.909	0.595	2.63	2.63	2.68	23.6	2.10	45.5	11.6	26.4
VW	0.80	0.0708	1.1	0.900	0.506	3.29	3.29	3.23	25.6	2.65	49.7	16.2	28.5
VW	0.80	0.0856	1.1	0.895	0.447	3.87	3.87	4.11	29.4	3.14	57.5	20.4	29.7
VW	1.60	0.0569	1.1	0.798	0.557	2.63	2.63	3.54	22.6	2.32	57.9	18.6	28.1
VW	1.60	0.0708	1.1	0.785	0.469	3.29	3.29	4.36	24.8	2.74	63.5	21.6	26.8
VW	1.60	0.0850	1.1	0.764	0.412	3.87	3.87	5.30	26.2	3.05	71.0	26.8	24.0
VW	2.40	0.0419	1.1	0.726	0.640	1.85	1.85	3.08	16.8	0.82	59.5	14.3	23.4
VW	2.40	0.0499	1.1	0.711	0.562	2.26	2.26	3.66	18.0	1.04	65.4	21.6	28.7
VW	2.40	0.0569	1.1	0.701	0.504	2.63	2.63	4.05	19.4	1.22	69.9	25.6	21.3
ST	1.17	0.0708	3.3	0.852	0.490	3.29	3.29	3.57	25.2	1.89	50.1	1.2	9.5
ST	1.17	0.0850	3.3	0.837	0.434	3.87	3.87	4.11	29.6	2.62	68.8	2.7	21.3
ST	1.17	0.1113	3.3	0.822	0.364	4.94	4.94	5.70	32.4	4.18	71.2	8.5	56.4
ST	1.17	0.1419	3.3	0.812	0.327	5.85	5.85	7.01	34.7	5.18	80.2	15.8	37.1
ST	1.97	0.0708	3.3	0.733	0.459	3.29	3.29	4.48	22.6	2.10	62.8	4.3	49.0
ST	1.97	0.0853	3.3	0.711	0.403	3.87	3.87	5.61	25.6	2.62	70.1	13.4	45.0
ST	1.97	0.0997	3.3	0.693	0.366	4.42	4.42	6.43	27.2	3.38	73.6	19.5	50.5
ST	1.97	0.1136	3.3	0.683	0.341	4.94	4.94	7.19	27.9	3.63	89.2	22.6	50.8
ST	2.37	0.0561	3.3	0.720	0.520	2.63	2.63	4.05	19.3	2.35	56.0	4.0	60.6
ST	2.37	0.0714	3.3	0.685	0.450	3.29	3.29	5.09	20.9	2.07	73.4	12.8	56.8
ST	2.37	0.0847	3.3	0.665	0.401	3.87	3.87	6.37	22.5	3.08	85.2	19.2	50.0
ST	2.37	0.0983	3.3	0.653	0.369	4.42	4.42	7.13	24.4	2.80	95.6	22.9	52.6
ST	3.23	0.0637	3.3	0.470	0.385	2.86	3.20	6.34	18.4	1.77	90.2	31.1	27.5
ST	3.23	0.0705	3.3	0.437	0.351	3.35	3.63	6.98	19.7	2.32	96.1	31.7	31.0
ST	3.23	0.0847	3.3	0.414	0.323	3.93	3.81	8.38	19.9	3.20	100.2	31.4	27.0
ST	3.55	0.0493	3.3	0.471	0.453	2.26	1.77	4.82	16.8	1.40	73.7	23.5	36.0
ST	3.55	0.0555	3.3	0.400	0.378	2.50	2.29	6.07	17.7	1.65	88.1	22.6	27.5
ST	3.55	0.0637	3.3	0.371	0.349	2.86	2.99	7.13	17.4	1.43	94.5	30.5	28.0
ST	3.55	0.0697	3.3	0.302	0.321	3.20	3.29	7.86	17.6	1.77	102.6	30.5	33.5
ST	3.66	0.0493	3.3	0.420	0.420	2.26	2.10	5.15	17.4	1.65	80.2	22.2	53.5
ST	3.66	0.0569	3.3	0.360	0.360	2.50	2.44	6.64	17.3	1.95	94.3	26.2	31.0
ST	3.66	0.0697	3.3	0.315	0.315	3.17	3.14	8.53	16.9	2.19	110.3	28.3	26.0
WW	1.61	0.0850	3.3	0.764		3.87	3.87	5.30	26.2			4.9	25.2
WW	1.61	0.0992	3.3	0.749		4.42	4.42	6.13	28.2			14.0	26.7
WW	1.61	0.1122	3.3	0.739		4.94	4.94	6.86	29.6			15.2	31.8
WW	2.22	0.0705	3.3	0.684		3.29	3.29	5.55	21.6			13.7	32.9
WW	2.22	0.0841	3.3	0.658		3.87	3.87	6.34	23.4			19.8	28.0
WW	2.22	0.0994	3.3	0.637		4.42	4.42	7.35	25.7			23.2	30.2

nearly conical. The effect of increasing the abutment length is shown in figure 17(c), with essentially the same discharge as in figure 17(a). The scour hole is obviously larger in both depth and volume. The point of deepest scour is again displaced laterally from the face of the abutment, and the scour hole is elongated as in figure 17(a).

Scour-hole contours for the longer vertical-wall abutment lengths (L_a/B_f = 0.88, 0.97, and 1.0) are illustrated in figure 18. The discharge of 0.0567 m^3/s is the same for all three abutment lengths in figure 18; however, the tailwater varies. It is equal to the normal floodplain depth (8.2 cm) in figure 18(b), and has a ratio to the normal floodplain depth of 1.95 and 2.32 in figures 18(a) and 18(c), respectively. The scour hole in figure 18(a) retains the laterally displaced elongated shape observed in figures 17(a) and 17(c); however, it extends into the sideslope of the main channel. As the abutment encroaches on the main channel in figures 18(b) and 18(c), there is a definite contraction-scour effect; however, multiple scour holes are apparent in figure 18(b) that combine local and contraction scour.

Scour-hole shapes for the spill-through abutment can be observed in figures 19 and 20. At the lower discharge of 0.0850 m^3/s in figure 19(a) as compared to figure 19(b), the scour hole is laterally displaced from the centerline of the abutment face. As the discharge increases, it moves to the upstream quadrant of the abutment face, and the deepest point of the scour hole is located immediately adjacent to the abutment. Comparing figure 19(b) with figure 18(c), for which the discharge and abutment length are essentially the same, the scour hole for the vertical-wall abutment is deeper and remains laterally displaced from the centerline of the abutment face. Apparently, all other things being equal, the discharge required to localize the scour hole to the upstream corner of the abutment is lower for the spill-through abutment. It is interesting to note from figure 19(c) that an increase in relative abutment length L_a/B_f to 0.88 does not change the essential shape of the scour hole. Although there is some minor encroachment of the scour hole into the main channel, it is much less than in the case of the vertical-wall abutment of the same length in figure 18(a), having a smaller discharge. However, comparing figures 19(c) and 20(a) for the same length of spill-through abutment (L_a/B_f = 0.88), but for a smaller discharge in figure 20(a), the scour hole moves almost completely into the main channel in figure 20(a). In figures 20(b) and 20(c), the scour process appears to be dominated by contraction scour that moves well upstream of the abutment as a headcut and extends considerably downstream as well.

The influence of the abutment shape at the same discharge and at essentially the same abutment length is illustrated in figure 21 where scour contours are shown for the vertical-wall, spill-through, and wingwall abutments. The maximum scour depths in figures 21(a), 21(b), and 21(c) are 29.0 cm, 22.9 cm, and 23.2 cm, respectively. The laterally displaced and elongated scour hole for the vertical-wall abutment shown in figure 21(a) moves next to the abutment face in the upstream quadrant in figure 21(b). In figure 21(c), the scour hole is also in the upstream quadrant of the abutment face; however, it is more conical in shape and has been displaced laterally. The dependence of the maximum scour depth on the abutment shape is related to the degree of flow blockage so that for long abutments with a high degree of blockage, the local effect of the separation zone on the scour as affected by shape diminishes.

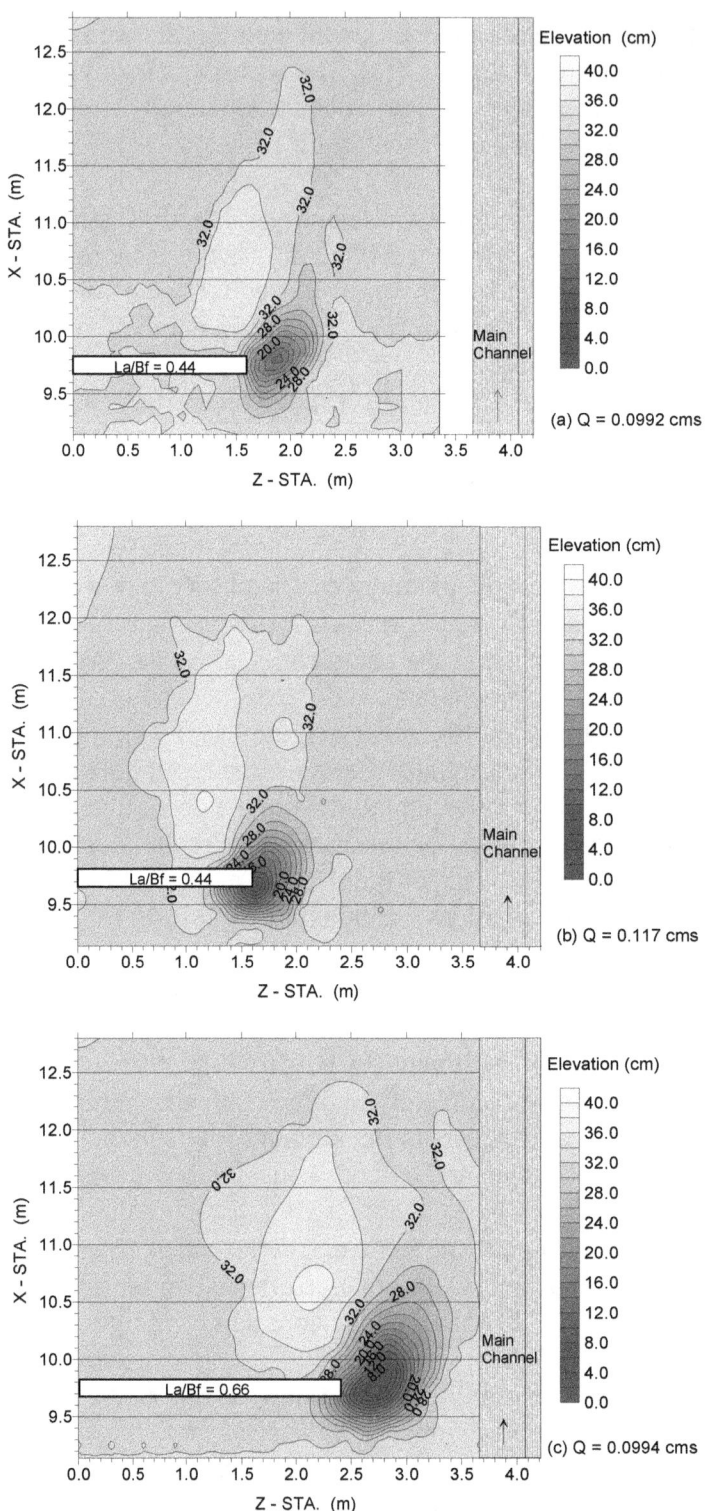

Figure 17. Bed elevations for shorter VW abutments after scour (cms = m³/s).

48

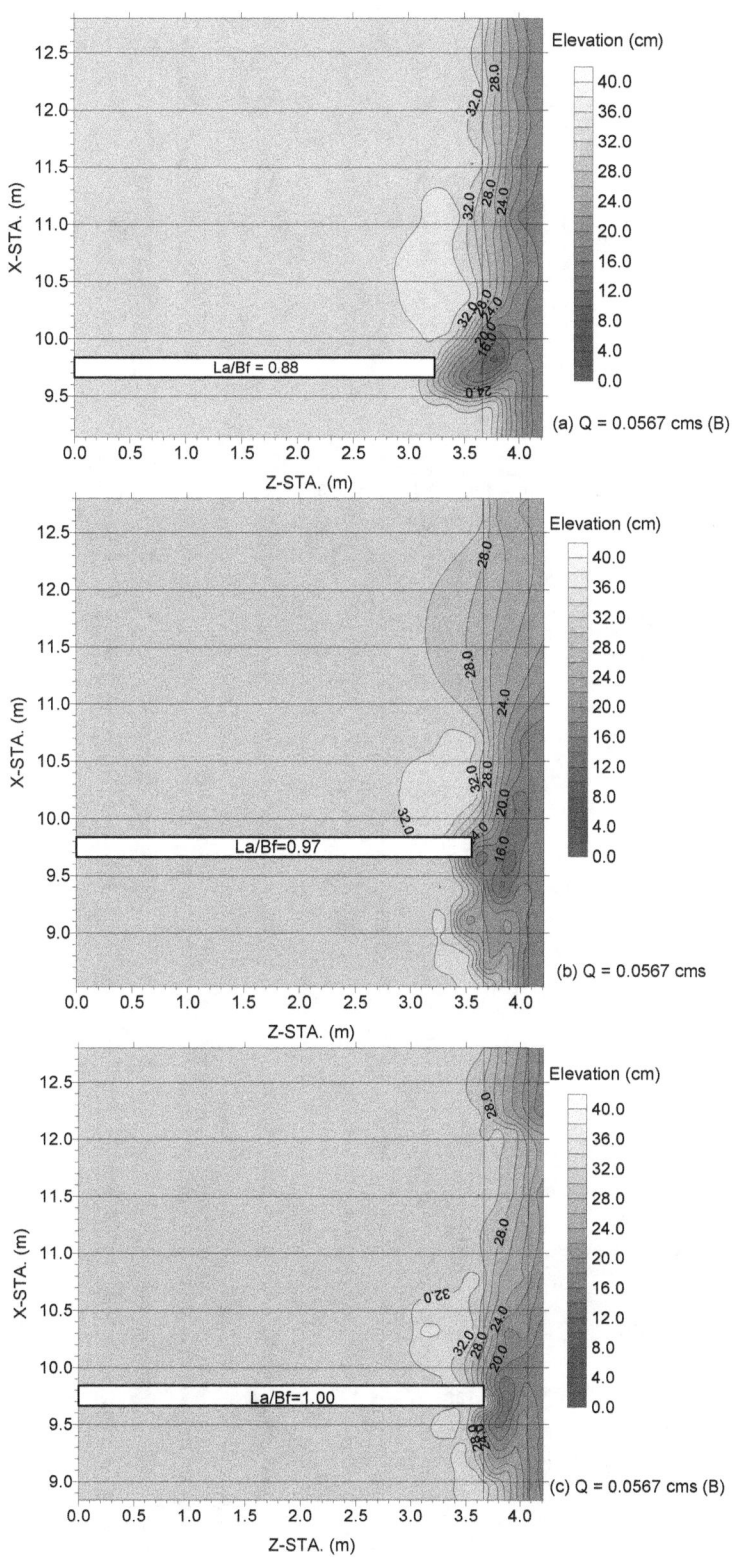

Figure 18. Bed elevations for longer VW abutments after scour (cms = m^3/s).

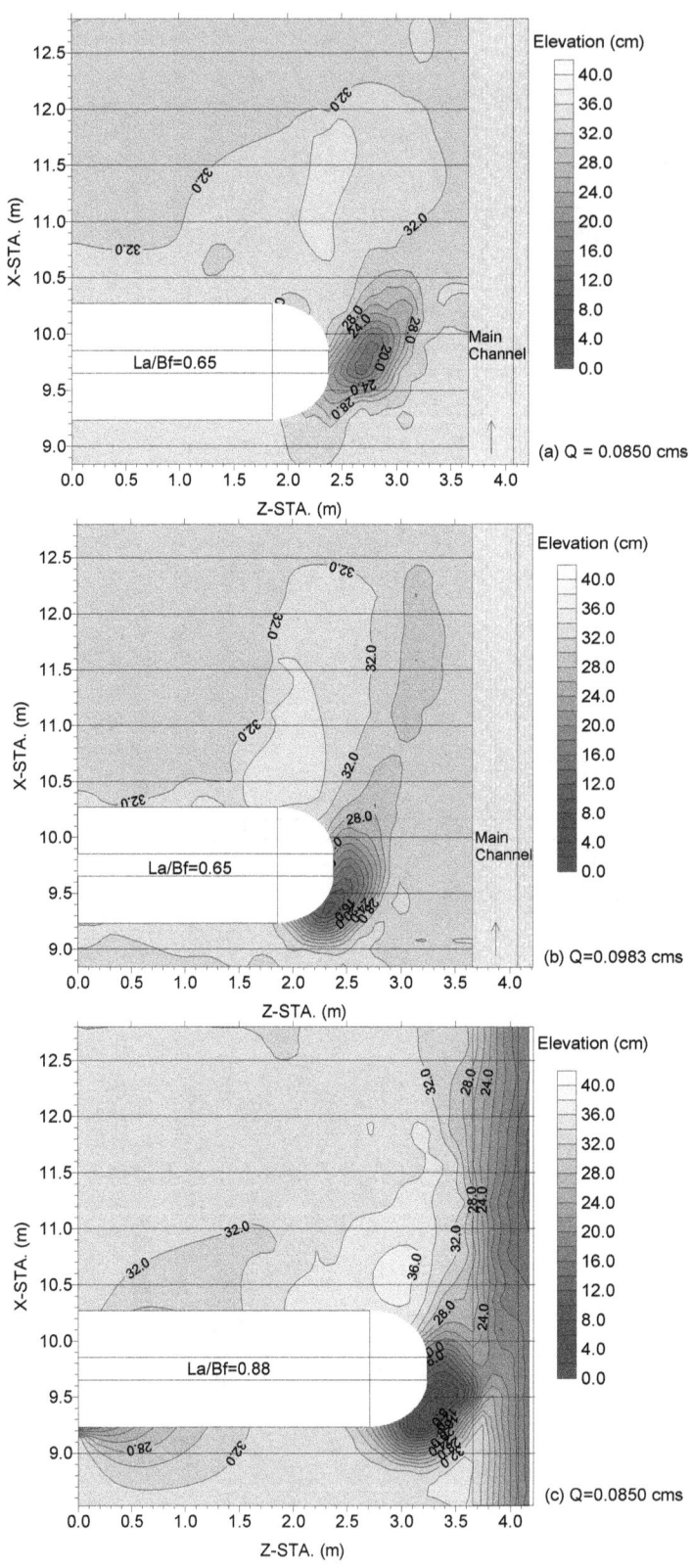

Figure 19. Bed elevations for shorter ST abutments after scour (cms = m³/s).

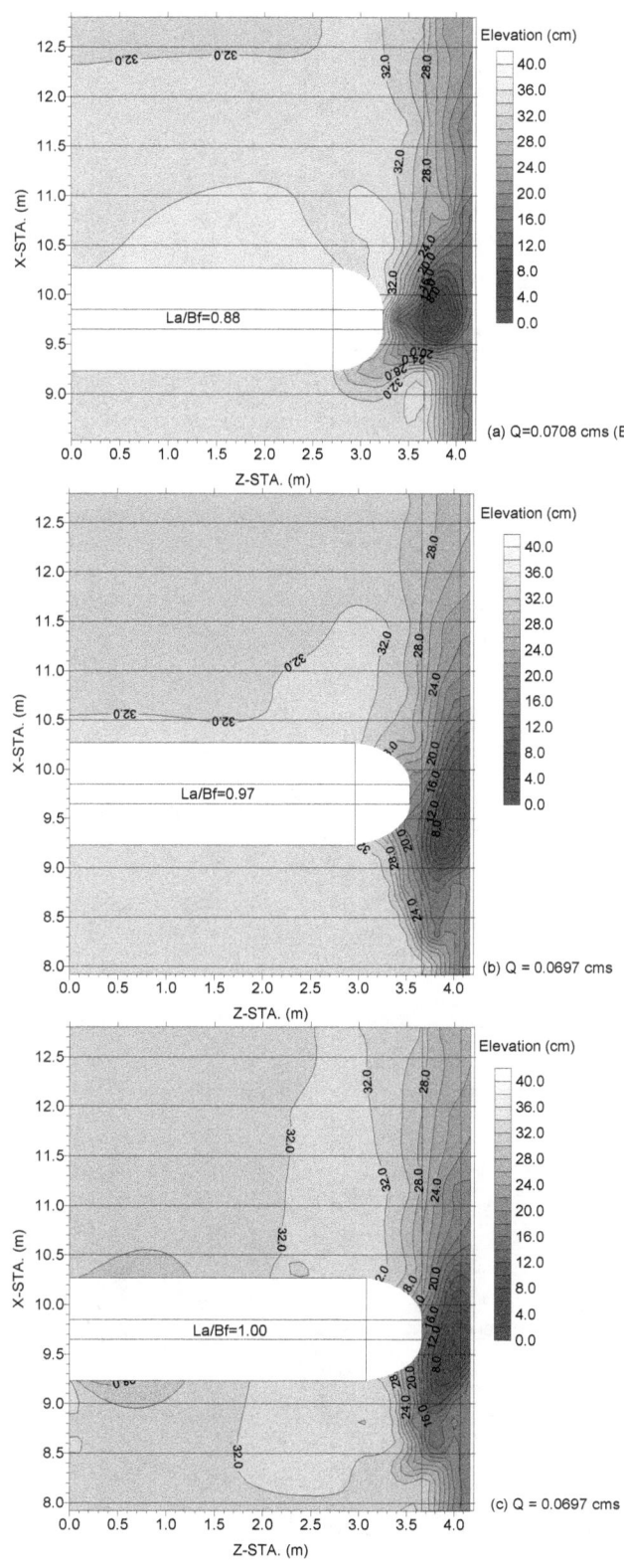

Figure 20. Bed elevations for longer ST abutments after scour (cms = m³/s).

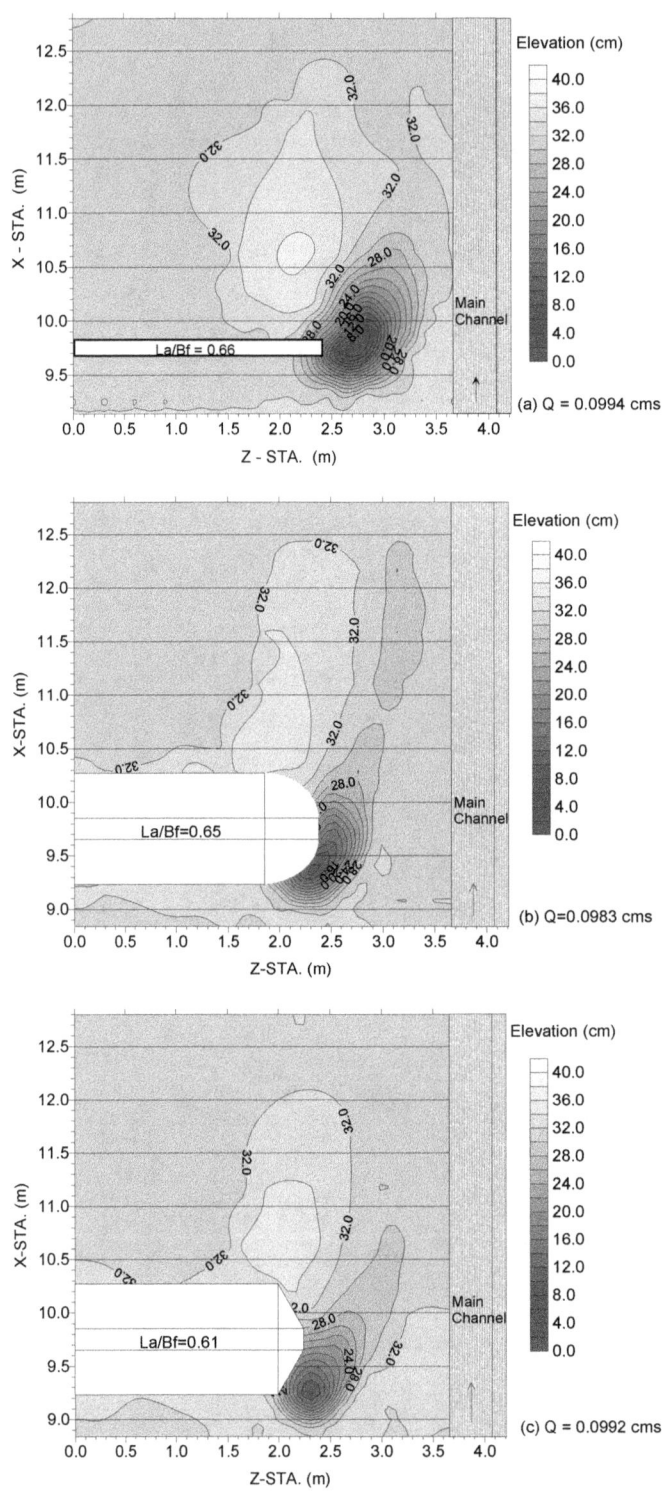

Figure 21. Bed elevations for VW, ST, and WW abutments after scour (cms = m³/s).

Sediment size effects on clear-water abutment scour are shown in figure 22 for sediments A, B, and C for the conditions of the same discharge and same length of a vertical-wall abutment. The scour-hole shapes and depths, as well as the downstream bar deposits for sediments A and B, are very similar (as shown in figures 22(a) and 22(b)), probably because of their small difference in size. For sediment C, however, the scour hole is significantly enlarged both in lateral extent and depth. In addition, the velocities exiting the main scour hole are sufficient to erode an elongated trench downstream of the main scour hole. There can be no doubt that sediment size as reflected by critical velocity has an important effect on the depth of the clear-water scour.

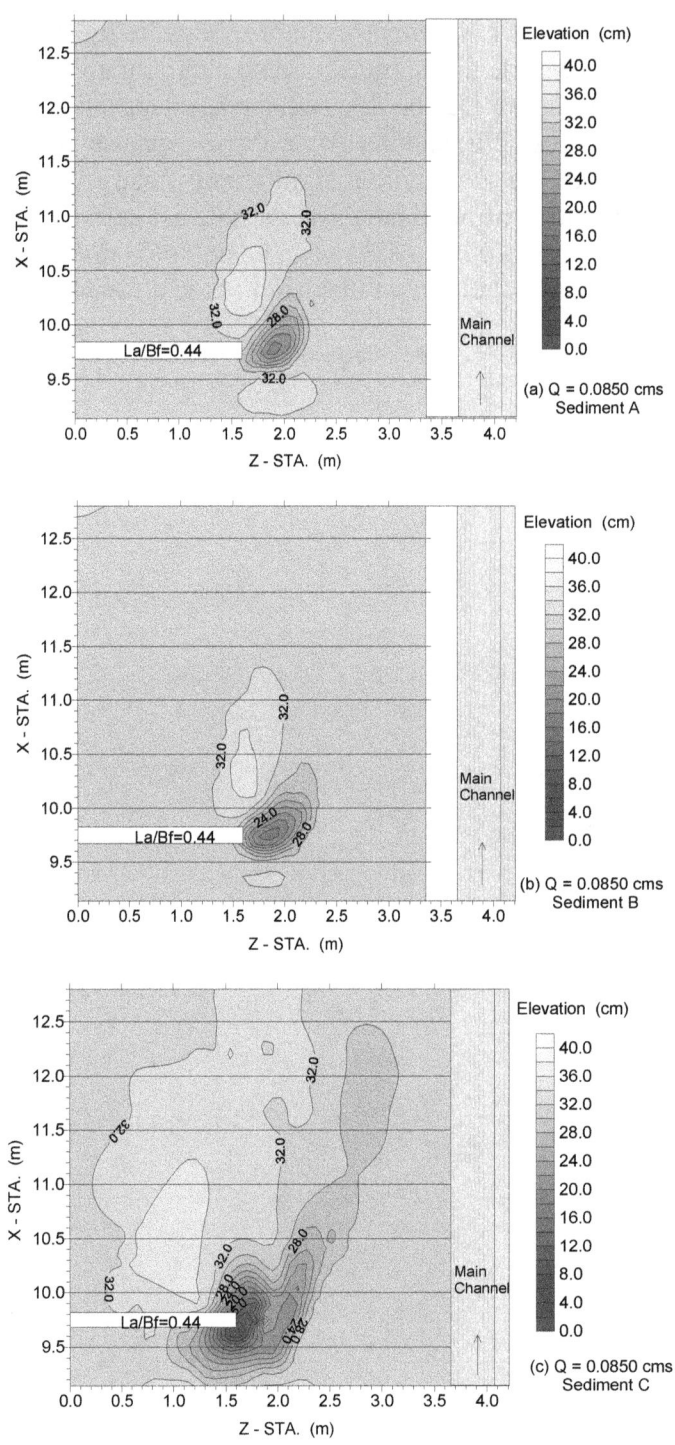

Figure 22. Bed elevations for VW abutments and sediments A, B, and C (cms = m³/s).

CHAPTER 4. ANALYSIS OF RESULTS

EQUILIBRIUM SCOUR DEPTH

Clear-water scour at a bridge abutment located on the floodplain of a compound open channel is influenced by the flow distribution between the floodplain and the main channel in the approach and contracted bridge cross sections. It has been shown that the effect of the flow contraction can be accounted for by the discharge contraction ratio M, which depends on both the abutment length and the discharge distribution in the approach section of a compound channel.[20] In addition, previous equations for predicting clear-water scour depths at abutments have depended on the ratio of floodplain velocity V_1 to critical velocity V_c in the bridge approach section,[14,20] with maximum clear-water scour occurring when V_1 approaches V_c (see chapter 2). In this formulation, the independent variables for scour prediction are determined from one-dimensional numerical models such as WSPRO. Sturm and Chrisochoides[86] have explored the suitability of these one-dimensional estimates of reference velocities and depths for scour prediction, and their results will be given in this chapter. For the case of significant backwater caused by the bridge opening, Sturm and Sadiq[23] have shown that the appropriate reference depth is the unconstricted floodplain depth rather than the approach floodplain depth with the bridge constriction in place. Extension of this methodology to abutments that encroach on the banks of the main channel in a compound channel has been reported by Sturm and Chrisochoides[87] and will be discussed in this chapter along with a consideration of the live-bed scour case.

A possible alternative to parameterizing scour depth in terms of approach velocity and discharge distribution is to relate it directly to local hydraulic conditions near the abutment face, although these conditions can only be predicted by two- or three-dimensional numerical models. Possible advantages of this alternative formulation include: (1) it may provide a means of unifying scour-prediction equations obtained from experiments in rectangular channels with those based on the more realistic compound-channel geometry, and (2) it may offer greater sensitivity of scour predictions to changes in local depth and velocity at the location of scour initiation near the abutment face. Theoretically, local scour is initiated when the ratio of local flow velocity to critical velocity exceeds unity, and it continues at an ever-decreasing rate until equilibrium is reached. The initial rate of scour and limiting equilibrium depth of scour might be expected to depend on the excess of the local velocity ratio in comparison to unity among other variables.

The implementation procedure for the prediction of clear-water abutment scour that is described herein is based on laboratory experiments in two different compound-channel cross sections as described in chapter 3. Abutment length, abutment shape, discharge, and sediment size were varied, and the resulting equilibrium scour depths were measured. In this chapter, a relationship for the prediction of equilibrium scour depth is formulated.

CLEAR-WATER SCOUR: FORMULATION I

The first formulation is based on Laursen's long contraction scour theory for clear-water scour[13] as modified by Sturm and Sadiq[23] for abutments ending on the floodplain of a compound channel (setback abutments). The long contraction theory is also extended to the case of the abutment encroaching on the bank of the main channel (bankline abutments) for both clear-water and live-bed scour. The theoretical development for clear-water scour is given first, followed by correlation of the scour data for both compound channels A and B.

Theoretical Clear-Water Contraction Scour for Setback Abutments

With reference to figure 23 for equilibrium scour conditions in a long contraction, it is assumed that the approach conditions tend toward the unconstricted depth and velocity on the floodplain, y_{f0} and V_{f0}, respectively. This assumption is consistent with the assumption of an idealized contraction with negligible head loss and velocity-head changes made by Laursen[13] in his analysis of clear-water contraction scour. The continuity equation for the floodplains from the approach section to the contracted section at equilibrium is:

$$V_{f0} \ y_{f0} \ = \ \mu_f \ V_{f2} \ y_{f2} \tag{24}$$

where V_{f0} and y_{f0} = average unconstricted velocity and depth in the approach floodplain, respectively; V_{f2} and y_{f2} = average velocity and depth in the contracted floodplain, respectively; and μ_f = generalized discharge contraction ratio. If the channel is rectangular with a width equal to the floodplain width, then $\mu_f = B_{f2}/B_{f0}$, the ratio of contracted floodplain width to approach floodplain width. However, in a compound channel, some of the approach floodplain flow joins the main-channel flow in the contracted section as shown previously in figure 7(b). Under these circumstances, μ_f becomes an empirical factor that depends on the flow distribution in the approach channel and its redistribution between the main channel and the floodplain in the contracted section. For equilibrium clear-water scour conditions, the velocity in the contracted floodplain section, V_{f2}, is set equal to the critical velocity, V_{f2c}. If this substitution is made in equation 24, and it is divided by the critical velocity in the approach section, V_{f0c}, the result is:

$$\frac{V_{f0}}{V_{f0c}} \ y_{f0} \ = \ \mu_f \ \frac{V_{f2c}}{V_{f0c}} \ y_{f2} \tag{25}$$

Now from equation 23 for critical velocity, the ratio of critical velocities, V_{f2c}/V_{f0c}, is proportional to the ratio of the corresponding depths to the 1/6 power (i.e., $(y_{f2}/y_{f0})^{1/6}$). If this substitution is made in equation 25 and it is solved for y_{f2}/y_{f0}, the resulting equation is:

$$\frac{y_{f2}}{y_{f0}} \ = \ \left[\frac{V_{f0} \ / V_{f0c}}{\mu_f}\right]^{6/7} \tag{26}$$

An equation of this form has been suggested by Richardson and Davis[8] in HEC-18 for estimating clear-water contraction scour, except that μ_f is evaluated as a geometric contraction ratio. Using the assumption that velocity-head changes and head losses at equilibrium contraction scour are small,[13] it can be shown that $y_{f2} = d_{sc} + y_{f0}$ (see figure 23). Then equation 26 can be written in terms of the theoretical contraction scour depth, d_{sc}, as:

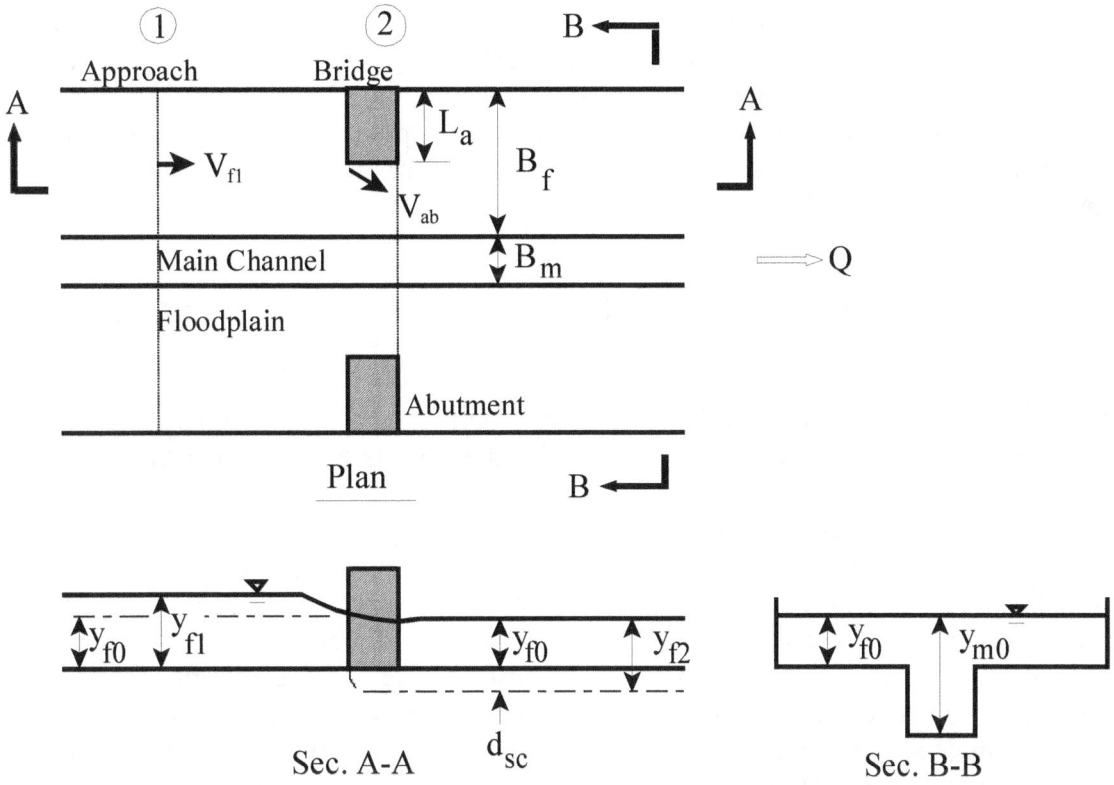

Figure 23. Definition sketch for idealized floodplain contraction scour
in a laboratory compound channel, compound channel A.

$$\frac{d_{sc}}{y_{f0}} + 1 = [\frac{V_{f0} / V_{f0c}}{\mu_f}]^{6/7} \qquad (27)$$

Laursen's[13] assumption is that local abutment scour $d_s = rd_{sc}$, where r is some constant greater than 1 that multiplies the theoretical contraction scour depth given by equation 27. Although the contraction scour predicted by equation 27 is for a fictitious long contraction with several restrictive assumptions, including the assumption that the approach depth and velocity tend toward their unconstricted values as scour approaches equilibrium, it suggests that an equation for abutment scour might take the form:

$$\frac{d_s}{y_{f0}} = C_r [\frac{V_{f0} / V_{f0c}}{\mu_f} - C_0] \qquad (28)$$

where C_r and C_0 are constants to be determined by experiment, and the exponent in equation 27 has been taken to be approximately unity as found in several experimental investigations.[14,19]

There are two dimensionless ratios of interest on the right-hand side of equation 28. The velocity ratio in the numerator is the unconstricted floodplain velocity in the approach section in ratio to the critical velocity under the same conditions. Theoretically, once this velocity ratio reaches the value of unity, maximum clear-water scour occurs and live-bed scour begins, assuming that the approach depth and velocity tend toward their unconstricted values as equilibrium is reached. The other independent dimensionless variable in equation 28, which is especially relevant in the present experiments, is the discharge contraction ratio at equilibrium, μ_f. Sturm and Janjua[20] suggested the use of the discharge contraction ratio M, which is defined as the ratio of unobstructed discharge in the approach channel to total discharge at the beginning of scour, as an estimate of μ_f. They showed that M was a good estimate of μ_f for small depth changes in the contraction, which was indeed the case for their experiments with a contraction experiencing negligible changes in depth through the contraction because the tailwater was high with respect to critical depth. However, if there is choking in the contraction and/or significant upstream backwater at the beginning of scour, then an improved estimate of μ_f is needed. If it is assumed that M is an estimate of q_{f1}/q_{f2} as shown by Sturm and Janjua,[20] then μ_f can be estimated as:

$$\mu_f = \frac{q_{f0}}{q_{f2}} = M \frac{q_{f0}}{q_{f1}} \qquad (29)$$

where M is evaluated for the approach depth at the beginning of scour and can be obtained from the WSPRO output,[21] $q_{f0} = V_{f0}y_{f0}$ at the end of scour as estimated by the unconstricted approach floodplain velocity and depth, and $q_{f1} = V_{f1}y_{f1}$ at the beginning of scour determined by the approach floodplain velocity and depth obtained from the constricted water-surface profile in the WSPRO output. Substituting equation 29 into equation 28 results in:

$$\frac{d_s}{y_{f0}} = C_r \left[\frac{q_{f1}}{M \, q_{f0c}} - C_0\right] \tag{30}$$

where $q_{f0c} = V_{f0c}y_{f0}$ and V_{f0c} is the critical velocity in the floodplain for the unconstricted floodplain depth y_{f0}. Equation 30 provides a working equation for fitting the experimental results.

Theoretical Clear-Water Contraction Scour for Bankline Abutments

For the abutment encroaching on the main-channel banks, the floodplain portion of the contracted section no longer exists, and the formulation of equation 30 has to be revised for this case. As shown in figure 24, the idealized contraction scour is assumed to occur in the contracted main-channel section, and by definition for $L_a = B_f$, M is exactly equal to q_{m1}/q_{m2} for a constant-width main channel. Thus, in the contracted section, continuity for critical conditions at equilibrium scour results in:

$$V_{m2c} y_{m2} = \frac{q_{m1}}{M} \tag{31}$$

where V_{m2c} = critical velocity in the main channel at the contracted section at equilibrium, y_{m2} = depth of flow in the contracted section in the main channel for equilibrium scour, q_{m1} = flow rate per unit width in the approach main-channel section at the beginning of scour, and M = discharge contraction ratio. Dividing equation 31 by $V_{m0c}y_{m0}$ gives:

$$\frac{V_{m2c}}{V_{m0c}} \frac{y_{m2}}{y_{m0}} = \frac{q_{m1}}{M \, V_{m0c} \, y_{m0}} \tag{32}$$

where V_{m0c} = the critical velocity in the main channel for the unconstricted flow depth in the main channel of y_{m0}. Using equation 23 to replace the ratio of critical velocities on the left side of equation 32 with the ratio of depths to the 1/6 power, $(y_{m2}/y_{m0})^{1/6}$, and solving for y_{m2}/y_{m0} produces:

$$\frac{y_{m2}}{y_{m0}} = \left[\frac{q_{m1}}{M \, (V_{m0c} \, y_{m0})}\right]^{6/7} \tag{33}$$

To be consistent with the case of the abutment terminating on the floodplain, it seems convenient to define the depth of scour relative to the floodplain elevation in both cases. Then, as

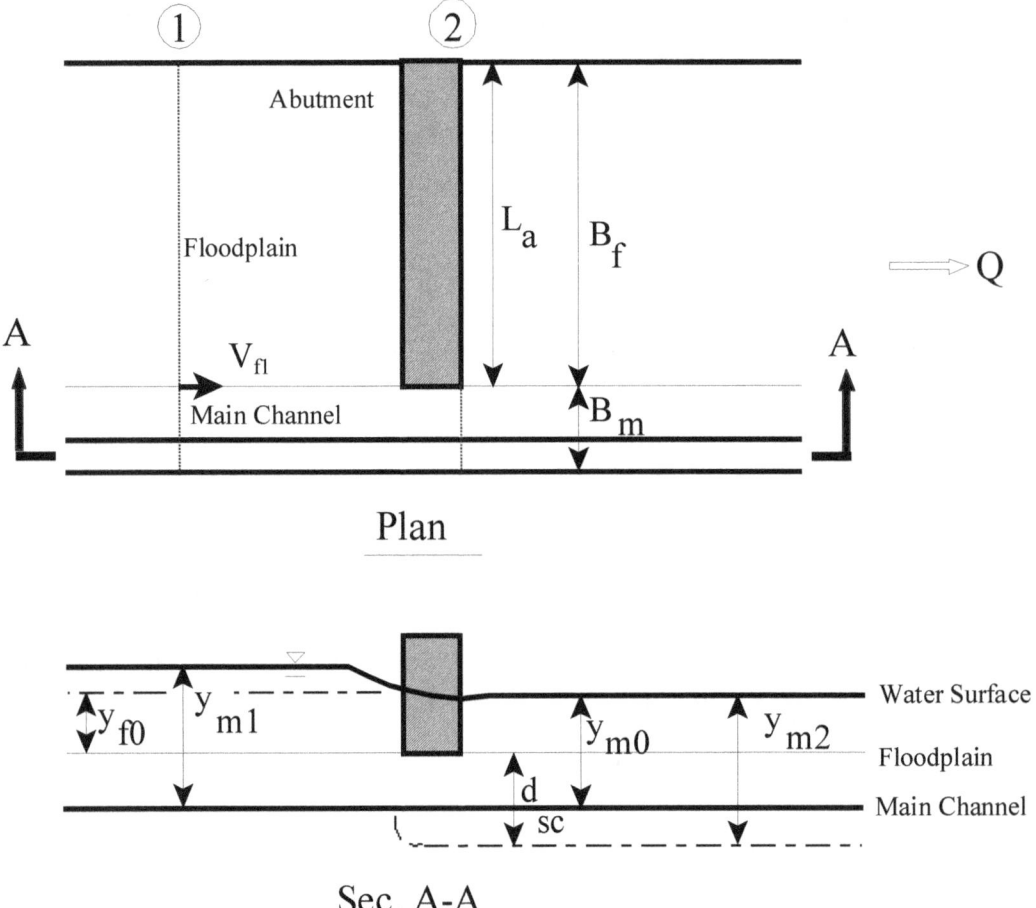

Figure 24. Definition sketch for idealized main-channel contraction scour in a laboratory compound channel, compound channel B.

shown in figure 24, $y_{m2} = d_{sc} + y_{f0}$. Again, for consistency with equation 30, equation 33 can be multiplied by y_{m0}/y_{f0} to yield:

$$\frac{d_{sc}}{y_{f0}} + 1 = \left[\frac{q_{m1}}{M(V_{m0c} \, y_{m0})} \right]^{6/7} \frac{y_{m0}}{y_{f0}} \tag{34}$$

Finally, if it is assumed that the 6/7 power can be approximated by unity as in the previous floodplain scour analysis, equation 34 can be placed in a form similar to equation 30:

$$\frac{d_s}{y_{f0}} = C_r' \left[\frac{q_{m1}}{M(V_{m0c} \, y_{f0})} - C_0' \right] \tag{35}$$

where the coefficients C_r' and C_0', in general, would not be the same for floodplain scour as given by equation 30.

Correlation of Results

The correlation of scour-depth measurements according to equation 30 is presented in figure 25 for the shorter abutment lengths with significant setbacks from the bank of the main channel (setback abutments). The results for vertical-wall abutments are given for both compound channels (A and B) and for relative abutment lengths L_a/B_f varying from 0.17 to 0.66. Sediment sizes of 1.1 mm, 2.7 mm, and 3.3 mm are included in the results in figure 25. Shown for comparison in figure 25 is the proposed relationship suggested by Sturm and Janjua[20] from a much more limited data set in a short, horizontal laboratory compound channel. In the latter data set, the tailwater essentially submerged the contraction so that there was very little backwater from the contraction; however, these data agree with the present data if the unconstricted floodplain depth is substituted for y_{f0}, even though it is not the normal depth. The best-fit relationship for all of the data is also given in the figure for which $C_r = 8.14$ and $C_0 = 0.40$ in equation 30. For this relationship, the coefficient of determination r^2 is 0.86, and the standard error of estimate in d_s/y_{f0} is 0.68.

In figure 26, the scour data for longer abutments that approach the bank of the main channel (bankline abutments) are correlated according to equation 35. These additional data comprise relative abutment lengths L_a/B_f of 0.88, 0.97, and 1.0. The major differences between equations 30 and 35 are the replacement of q_{f1} in equation 30 with q_{m1} in equation 35, and the evaluation of the critical velocity in the main channel for the unconstricted depth y_{m0} in equation 35 instead of the critical velocity in the floodplain in equation 30. The value q_{m1} was evaluated approximately as the measured discharge in the approach main channel divided by the main-channel topwidth. For those cases in which the tailwater depth exceeded normal flow depth, the tailwater depth y_{ftw} was substituted for the normal depth y_{f0}. It can be observed from figures 11 and 12 that the tailwater depth is representative of the depth in the contracted section as was the normal depth

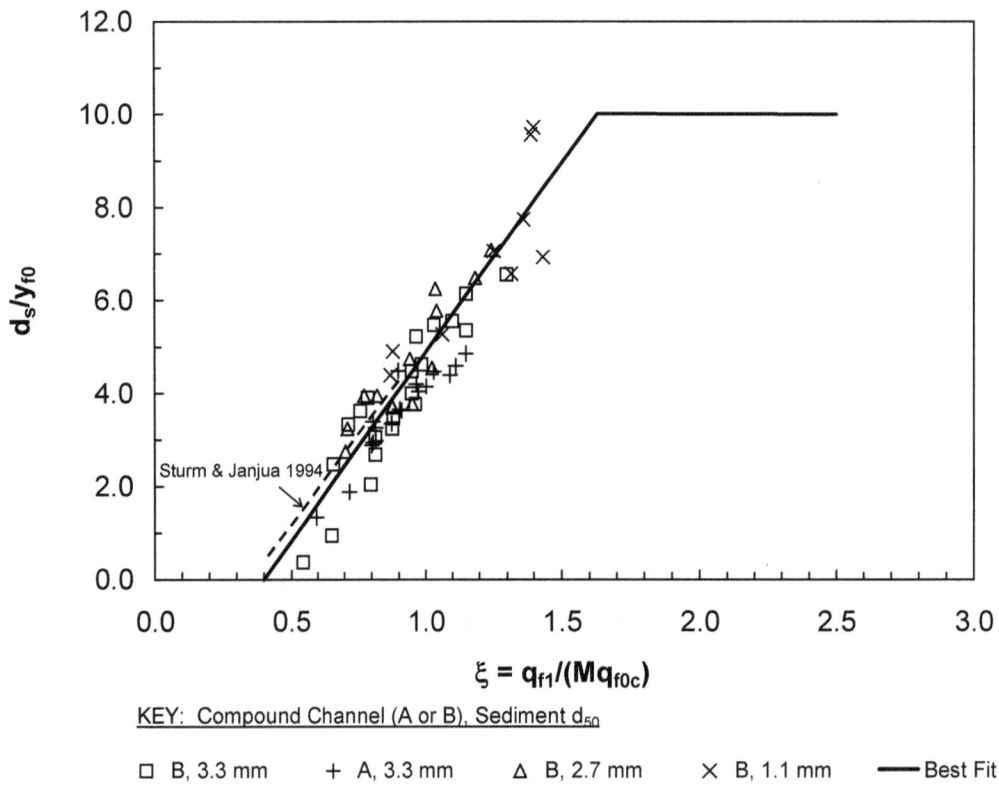

Figure 25. Scour-depth relationship based on approach hydraulic variables for VW abutments with $0.17 \leq L_a/B_f \leq 0.66$.

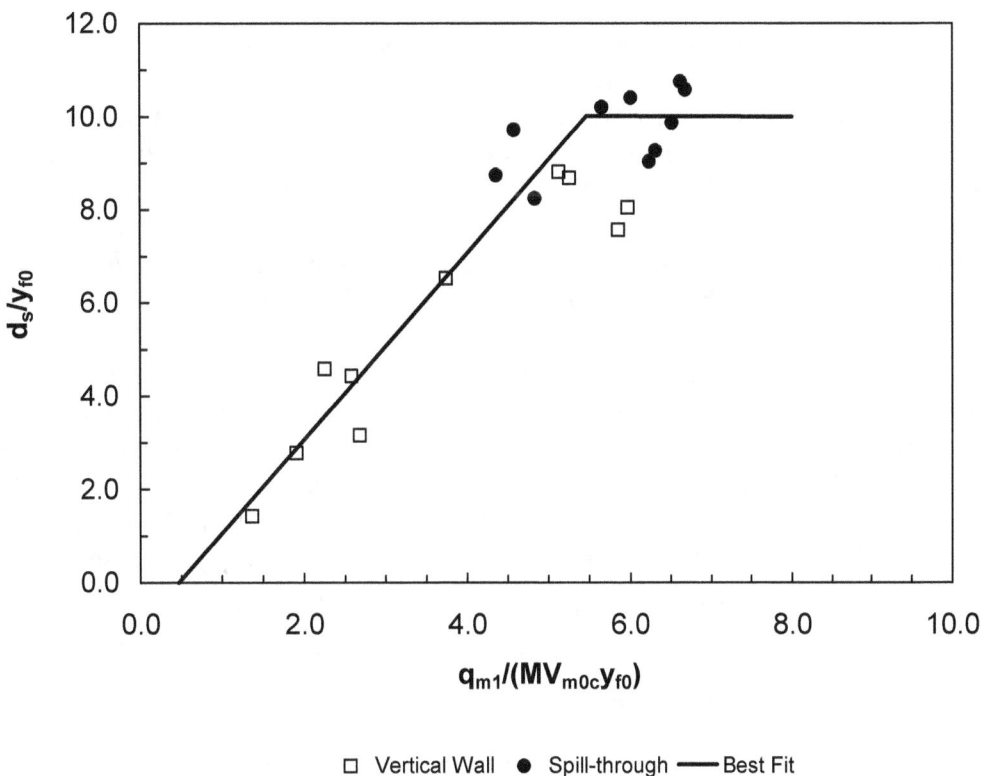

Figure 26. Scour-depth relationship based on approach hydraulic variables in main channel for VW and ST abutments with $0.88 \leq L_a/B_f \leq 1.0$ and $d_{50} = 3.3$ mm in compound channel B.

for the unsubmerged cases. Strictly speaking, the derivation of equation 35 applies only to the case of $L_a/B_f = 1.0$; however, the additional abutment length data seem to follow the same trend in figure 26. The scour hole clearly extends from the floodplain into the main channel for $L_a/B_f = 0.88$ in figures 18(a) and 20(a), while it is localized near the abutment face on the floodplain in figure 19(c). In all cases of $L_a/B_f = 0.97$ and 1.0 in figures 18, 19, and 20, the scour hole is located in the main channel. The best-fit values of the coefficients in equation 35 are $C_r' = 2.0$ and $C_0' = 0.47$, with an $r^2 = 0.92$ and a maximum value of d_s/y_{f0} of approximately 10.

The data shown in figure 26 represent the maximum total scour depths regardless of their classification as local or contraction scour. Indeed, these two types of scour occur simultaneously at different locations, as in figure 18(b), while interfering and combining as in figure 18(c), for example. Furthermore, the lengthening of the scour hole in the upstream direction in figures 20(b) and 20(c) seems to reflect the interaction of the floodplain and main-channel flows well upstream of the abutment in a contraction-scour effect. *Based on these results, the artificial addition of contraction and local abutment scour when the abutment is near the bank of the main channel may be responsible for the overestimates of scour in these cases.*

If the scour is a combined local and contraction effect for abutments near the bank of the main channel, it would seem that either equation 30 or 35 might serve as scour predictors because both are based on acceleration of the flow caused by the contraction (i.e., the floodplain flow in the case of equation 30 and the main-channel flow in the case of equation 35). Accordingly, the data for both short and long abutments are combined in figure 27, with the critical velocity calculated in the main channel for $L_a/B_f = 0.97$ and 1.0, and in the floodplain for all other cases. In other words, equation 30 is modified as:

$$\frac{d_s}{y_{f0}} = C_r \left(\frac{q_{f1}}{M V_{xc} y_{f0}} - C_0 \right) \tag{36}$$

where $V_{xc} = V_{f0c}$ for abutments located on the floodplain (setback abutments), and $V_{xc} = V_{m0c}$ for abutments near the bank of the main channel (bankline abutments). In addition, the data for spill-through and wingwall abutments are incorporated into figure 27. It is apparent that the relative scour depth levels off to a constant value for large values of the independent variable just as in figure 26. The actual mean value of the relative scour depth is 9.4 ±1.0 after the point of leveling off; however, considering the uncertainty involved and remaining on the conservative side, the maximum value is shown as 10 in figure 27. The scour data for the linear portion of the graph are compared with the best-fit relationship given previously in figure 25 for the setback abutments only. The data for compound channel A are not shown in the figure to retain clarity; nevertheless, the best-fit regression analysis remains essentially unchanged with the addition of the long abutment data ($C_r = 8.14$, $C_0 = 0.40$), except that the standard error of estimate for the relative scour depth increases from 0.68 to 0.75. In all cases, the scour depth is defined as the maximum depth in the scour hole measured below the undisturbed floodplain elevation. The fact that equations 30 and 35 both predict the scour depth reasonably well for the long abutments seems to be indicative of the contraction process controlling scour, whether it is measured by the main-channel flow acceleration or the floodplain flow acceleration. Both equations 30 and 35

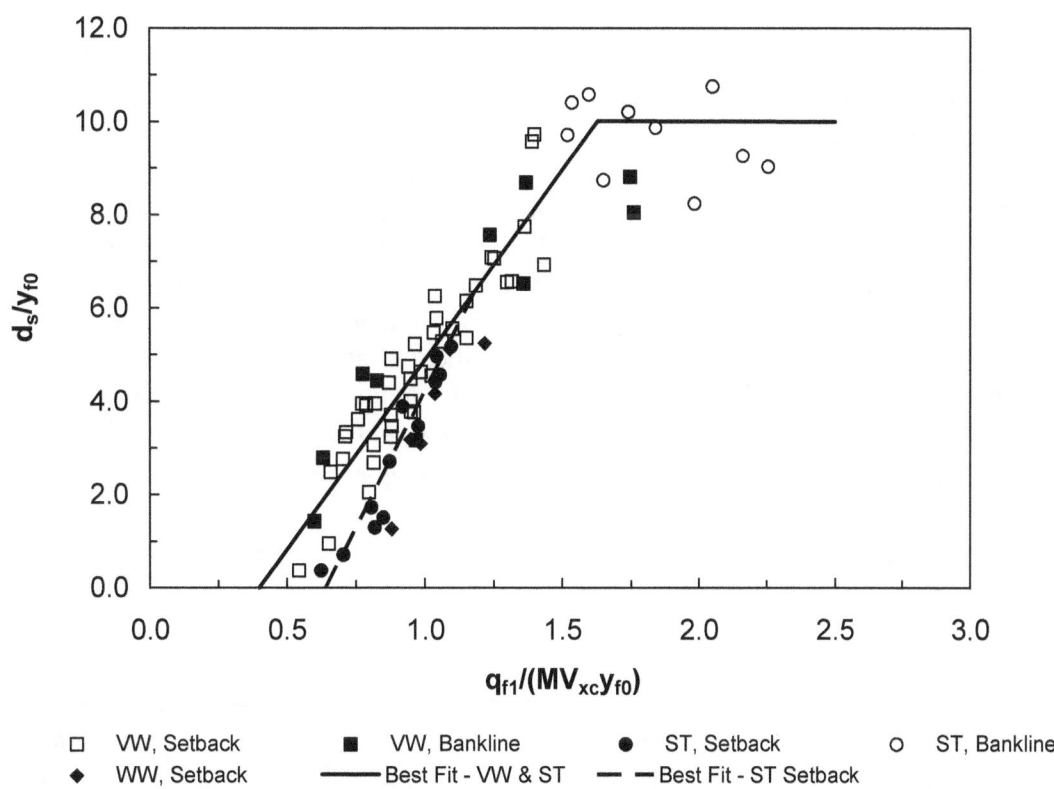

Figure 27. Scour-depth relationship based on approach hydraulic variables
in floodplain for all abutments and sediments, compound channel B.

require an estimate of average floodplain depth in natural channels, and it is recommended that the average depth of flow blocked by the abutment and embankment on the floodplain in the approach section be used for the unconstricted flow (y_{f0}) and the constricted flow (y_{f1}) for consistency.

Also shown in figure 27 are data for the spill-through and wingwall abutments. The spill-through data appear in two different regions, depending on the magnitude of the independent variable in figure 27. For shorter abutments, or more precisely, abutments that block less floodplain flow, the scour depth is smaller for the spill-through than for the vertical-wall abutment. The difference becomes less and less as the independent variable increases, which corresponds to an increasing discharge or an abutment that blocks more flow. The spill-through data seem to merge with the vertical-wall data for $q_{f1}/(MV_{xc}y_{f0})$ greater than about 1.2 and remain comingled with the vertical-wall data as the best-fit line levels off at a maximum value of relative scour depth of about 10. Apparently, the local flow effects at the face of the abutment caused by the differences in the abutment shape become unimportant as the degree of contraction becomes large. A scour-depth correction for the spill-through shape can be derived from the data in figure 27. A least-squares best fit for the spill-through data produces a linear relationship, as shown in figure 27, having a coefficient of determination of $r^2 = 0.93$, with a standard error of estimate for d_s/y_{f0} of 0.45. If the relative scour depth for a spill-through abutment is taken as a ratio to the value for a vertical-wall abutment, the result is a shape correction factor for the spill-through abutment K_{ST}:

$$K_{ST} = 1.52 \; \frac{\xi - 0.67}{\xi - 0.40} \qquad for \;\; 0.67 \leq \xi \leq 1.2 \qquad (37)$$

where $\xi = q_{f1}/(MV_{xc}y_{f0})$, and $K_{ST} = 1.0$ for $\xi \geq 1.2$ and 0 for $\xi \leq 0.67$. The shape factor of 0.55 given in HEC-18[8] for spill-through abutments corresponds to a ξ value of approximately 0.82, which would correspond to short setback abutments for which M approaches 1, $y_{f0} = y_{f1}$, and $V_{f1}/V_c = 0.82$. In other words, the shape factor of 0.55 given in HEC-18 should be viewed as an average value for short abutments with no backwater and conditions approaching threshold live-bed scour. Melville[24] has also observed that the shape factor approaches a value of 1 for very long abutments, although his values are given in terms of the abutment length rather than the discharge contraction ratio. The wingwall data shown in figure 27 are not significantly different from the spill-through data. This is a result of the fact that the approach velocity and depth were not measured for this case, but rather estimated from the vertical-wall abutment data. Although there are only small differences in M for the spill-through versus vertical-wall abutment data in figure 7(a), apparently the combined differences in approach depth, velocity, and M for the wingwall abutment are sufficient to obscure the shape effects when the independent variables are estimated rather than measured. Thus, the wingwall shape data are inconclusive.

Because of the influence of the discharge contraction ratio M and the backwater effects, it should not be assumed that the maximum clear-water scour depth of 10 times the unconstricted floodplain depth as shown in figure 27 is always the appropriate estimate of equilibrium scour depth. In other words, the scour depth can be less than 10 times the unconstricted floodplain flow depth measured below the floodplain even though the approach velocity has reached critical velocity and the threshold of live-bed scour. For example, as the abutment length decreases and less flow is blocked, M increases and approaches 1. In addition, backwater effects become less likely so that y_{f1} approaches y_{f0}. So even if $V_{f1} = V_c$, equation 29 gives a maximum relative scour depth d_s/y_{f0} of approximately 5. Considered from a different viewpoint, large main channels with very rough floodplains tend to have larger values of M, which reduce the relative scour depth even if threshold live-bed conditions are reached in the approach channel. In addition, for increasing tailwater depths at the same discharge and abutment length, or for increasing the abutment length at the same discharge, the approach floodplain velocity is reduced, making threshold live-bed scour conditions much less likely to occur. In conclusion, *it should not be assumed from figure 27 that a scour depth of 10 times the unconstricted floodplain depth is always the maximum equilibrium scour depth expected.*

CLEAR-WATER SCOUR: FORMULATION II

Clear-water scour at a bridge abutment located on the floodplain of a compound open channel occurs at an ever-decreasing rate from the initiation of scour until equilibrium is achieved. Theoretically, local scour is initiated when the ratio of the local bed shear velocity U_* to its critical value U_{*c}, or the ratio of local flow velocity V to critical velocity V_c, exceeds unity. Furthermore, the initial rate of scour and limiting depth of scour have been shown to increase with the value of V/U_{*c} in experiments on scour by jets of water.[79] Thus, although the local depth-averaged velocity near the abutment face is continually decreasing as the scour hole grows larger with time, it seems possible to relate the maximum depth of scour to the maximum depth-averaged velocity near the abutment face V_{ab} at the beginning of scour as defined in figure 23. This velocity cannot be predicted by one-dimensional models. Biglari and Sturm[55] have applied a depth-averaged k-ε turbulence model to predict the flow field and V_{ab} around an abutment on the floodplain of a compound channel using compound channel A and vertical-wall abutments. For compound channel B, V_{ab} was measured for several discharges for each abutment length.

Dimensional Analysis

As a test of the hypothesis that scour depends on the local depth-averaged velocity near the abutment face, a relationship for the equilibrium clear-water scour depth is sought in the form:

$$d_s = f [y_{f0}, y_{ab}, V_{ab}, \rho, (\rho_s - \rho), g, d_{50}] \qquad (38)$$

where d_s = equilibrium scour depth, y_{f0} = unconstricted flow depth in the floodplain set by the uniform flow downstream of the bridge, y_{ab} = floodplain flow depth at the location of V_{ab} in the contracted section, V_{ab} = maximum velocity near the upstream corner of the abutment face, ρ = fluid density, ρ_s = sediment density, g = gravitational acceleration, and d_{50} = median sediment diameter. Dimensional analysis of equation 38 results in:

$$\frac{d_s}{y_{f0}} = f[\frac{y_{ab}}{y_{f0}}, N_s, F_{ab}, \frac{d_{50}}{y_{ab}}] \qquad (39)$$

where $F_{ab} = V_{ab}/(gy_{ab})^{0.5}$ = Froude number in the contracted floodplain near the abutment face, and $N_s = V_{ab}/[(SG - 1)gd_{50}]^{0.5}$ = sediment number in the contracted section as defined by

Carstens,[80] with SG = specific gravity of the sediment. The value of the relative roughness (d_{50}/y_{ab}) can be replaced by the critical value of the Froude number F_c, or the critical value of the sediment number N_{sc}, for the case of fully rough turbulent flow[85] as in these experiments. For quartz sediment and water, either F_{ab} or N_s can be considered redundant with respect to the other. If it is further assumed that scour is related to the excess of velocity (or sediment number) with respect to its critical value as suggested by several sediment transport formulas,[80, 90-92] a relationship for dimensionless scour can be presented as:

$$\frac{d_s}{y_{f0}} = f\left[N_s - N_{sc}, \frac{y_{ab}}{y_{f0}} \right] \tag{40}$$

Alternatively, the scour relationship could be given as:

$$\frac{d_s}{y_{f0}} = f\left[\frac{V_{ab}}{V_c} - 1, \frac{y_{ab}}{y_{f0}} \right] \tag{41}$$

Correlation of Results

Figure 28 illustrates the correlation of scour data in terms of the excess velocity ratio suggested by equation 41. The experimental scour data for compound channel A were measured by Sadiq[22] and reported by Sturm and Sadiq,[23] while the scour data for compound channel B were collected as part of this study. The values of V_{ab} and y_{ab} were predicted by Biglari's numerical two-dimensional turbulence model for compound channel A,[54] while they were measured for compound channel B. For compound channel A, the left and right floodplains were slightly asymmetrical because of finite construction tolerances so that slightly different scour depths were measured for the left and right abutments. These two scour depths in each experiment were averaged for comparison with the numerical model runs, which assumed perfect symmetry. The value of V_c was calculated from Keulegan's equation as a function of d_{50}/y_{ab}, with the appropriate values of Shields' parameter.

The results in figure 28 include experimental data for compound channels A and B for six different values of L_a/B_f up to a maximum of 0.66, and for three different sediment sizes (1.13 to 3.30 mm). The value of L_a/y_{f0} varied from 3 to 90, which includes both intermediate- and large-scale abutment lengths according to Melville's classification.[14] The variable y_{ab}/y_{f0} in equation 41 is an indication of relative local water-surface drawdown near the upstream corner of the abutment; however, it was found not to be a significant explanatory variable for the observed scour depths in these experiments. The coefficient of determination for the least-squares best-fit relationship in figure 28 is $r^2 = 0.88$, and the best-fit equation is:

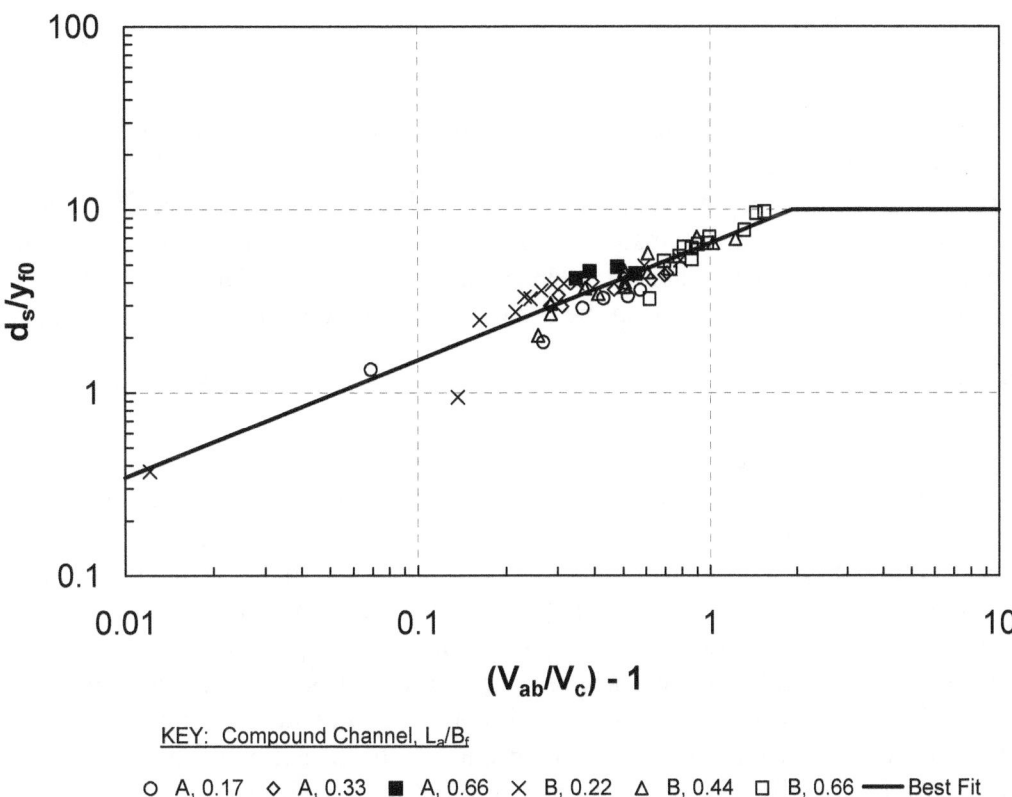

Figure 28. Scour-depth relationship based on local hydraulic variables for
VW abutments with $L_a/B_f \leq 0.66$ and d_{50} = 3.3, 2.7, and 1.1 mm.

$$\frac{d_s}{y_{f0}} = 6.55 \left[\frac{V_{ab}}{V_c} - 1 \right]^{0.64} \tag{42}$$

for $[(V_{ab}/V_c) - 1] \leq 1.94$ with $d_s/y_{f0} = 10$ for greater values. Correlation of the data according to the excess sediment number as given in equation 40 was not as successful as the result given by equation 42.

If the data for longer abutments are also plotted according to equation 42, it can be observed in figure 29 that the best-fit relationship does not compare well with these data. It is plausible that the measurement of V_{ab} as a depth-averaged, resultant velocity is no longer sufficient to explain the complex, three-dimensional flow pattern that occurs at the face of the abutment as the abutment encroaches on the main channel. Clearly, an additional flow factor is needed to explain better the scour data for this case, barring the use of a three-dimensional numerical model.

DISCUSSION

A comparison of figures 25 and 28 shows relatively little difference in using formulations I or II for scour prediction for relative abutment lengths less than 0.66. Formulation I is based on variables easily determined by WSPRO, while formulation II requires output from at least a two-dimensional, depth-averaged numerical model. It is not clear that the finite elements surface water modeling system (FESWMS) will satisfy this need with its assumption of constant eddy viscosity. Biglari's model[54-55] used a k-ε turbulence closure to capture the interaction between the main-channel and floodplain flow, and even higher order turbulence closure models may be needed to model the large-scale eddies of separation at the abutment. Only three-dimensional numerical models can be used to even attempt a prediction of the action of the horseshoe vortex and its effect on scour depths. This seems to be borne out by the relatively large deviation in the scour data for longer abutments in figure 29 from the best-fit relationship obtained for setback abutments that terminate in the floodplain well back from the main channel. Nevertheless, formulation II is attractive because of the expected rapid development and use of two-dimensional and three-dimensional numerical models in the next few years, and also because of the possibility of unifying the experimental results on abutment scour from rectangular channels with those from compound channels.

There is ample precedent for taking the "hydraulic approach" and developing a simple method for predicting V_{ab} along with a velocity adjustment factor for longer abutments until more advanced numerical models are readily available and usable. For example, one possible reference velocity is the mean cross-sectional velocity in the contracted cross section that can be obtained from WSPRO. However, relating V_{ab} to the average cross-sectional velocity for the whole cross section is complicated by the fact that WSPRO predicts the average cross-sectional velocity at the downstream face of the bridge, whereas V_{ab} is the *resultant, depth-averaged velocity* at the upstream corner of the abutment face. In addition, the average cross-sectional velocity is not at all representative of the floodplain velocity for setback abutments that are located on the

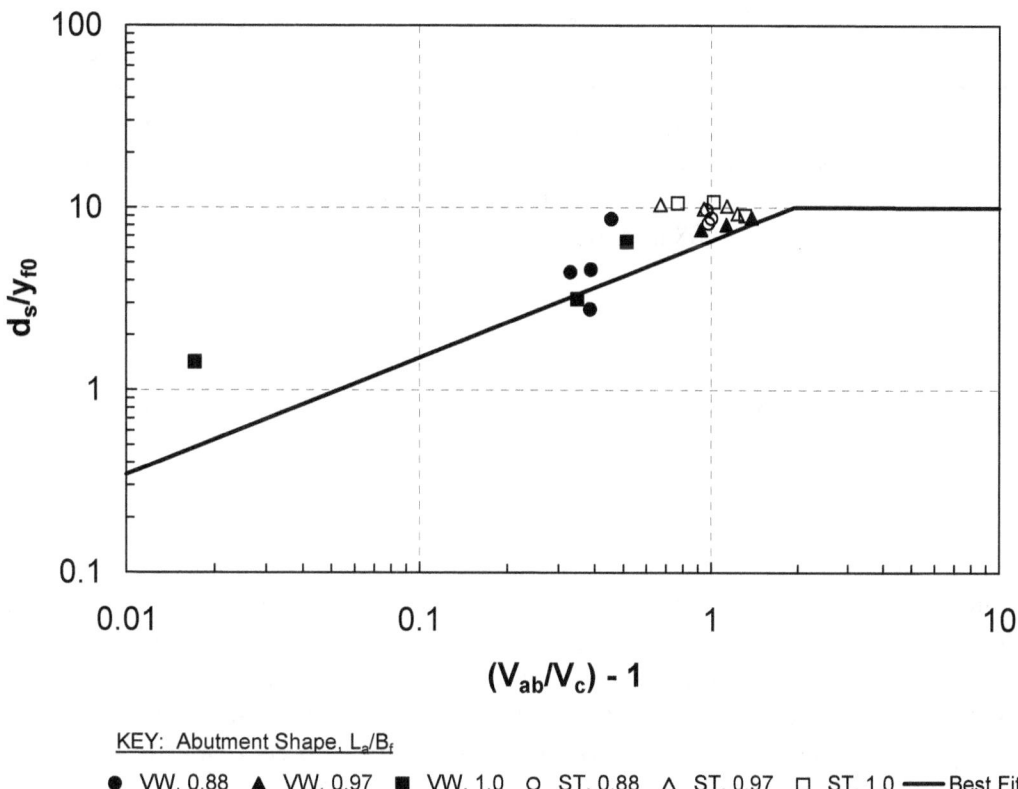

Figure 29. Scour-depth relationship based on local hydraulic variables for VW and ST abutments with $L_a/B_f \geq 0.88$ and $d_{50} = 3.3$ mm.

floodplain. The floodplain velocity in the contracted section is, in reality, determined by both the local acceleration near the abutment face and the entrainment of the floodplain flow into the main channel. Furthermore, the practice of taking the WSPRO streamtube velocity nearest the abutment face in the floodplain is bound to fail (as will be shown subsequently) because the flow distribution in WSPRO is determined by the conveyance distribution, which does not reflect the flow acceleration around the abutment face. As the abutment encroaches on the main channel, the entrainment of the floodplain flow into the main-channel flow occurs near the abutment face so that again a simple estimate of mean cross-sectional velocity cannot properly predict the scour potential in this highly three-dimensional flow region. In spite of these difficulties, there remains one other possibility suggested by the fact that the independent dimensionless variables in figures 25 and 28 provide a similar degree of explanation of scour depth so that they must be related to each other. The correlation between the two for the shorter abutments ($L_a/B_f \leq 0.66$) is shown in figure 30(a), with $r^2 = 0.88$ and the best-fit equation as:

$$\frac{V_{ab}}{V_c} - 1 = 1.56 \left(\frac{q_{f1}}{M \, q_{f0c}} - 0.4 \right)^{1.78}$$
(43)

Equation 43 has the advantage that it depends on hydraulic variables on the right-hand side that can be predicted well by a one-dimensional model such as WSPRO. However, the data for longer abutments ($L_a/B_f \geq 0.88$) are plotted in figure 30(b), and the correlation of figure 30(a) no longer holds. This is the reason for the disagreement between measured and predicted scour in figure 29, where the measured $[(V_{ab}/V_c) - 1]$ is used as the independent variable for the longer abutments. The resultant velocity was measured near the upstream corner of the abutment face for both the shorter and longer abutments. In the latter case, it is likely that a significant downward vertical component of the velocity existed that was not measured by the miniature propeller meter, and so the resultant velocity was underestimated. Only a three-dimensional laser Doppler velocimeter (LDV) with a fiber-optic probe located inside a Plexiglas abutment could be used to measure the resultant velocity near the abutment in these experiments, or it could be computed using a verified three-dimensional numerical model. Both of these methods were outside the scope of the present study. *In conclusion, formulation I is recommended as the best overall clear-water scour-prediction method for both short and long abutments.* The limitations of the recommended scour-prediction equation are discussed at the end of the chapter.

APPLICATION TO LIVE-BED CONDITIONS

Live-bed scour is the condition in which sediment transport occurs in the channel approaching the bridge contraction so that sediment continuously passes through the scour hole. Equilibrium is reached when the sediment transport rate out of the scour hole equals the sediment transport rate into the scour hole. For a compound channel in riverbank flow, the floodplain velocities are quite likely to be less than the critical velocity so that live-bed conditions can be realized only in the main channel. Thus, as a practical matter, abutment scour under live-bed conditions is often limited to the case of the abutment's encroaching on the bank of the main channel with sediment transport occurring only in the approach main channel. Even this condition may be difficult to

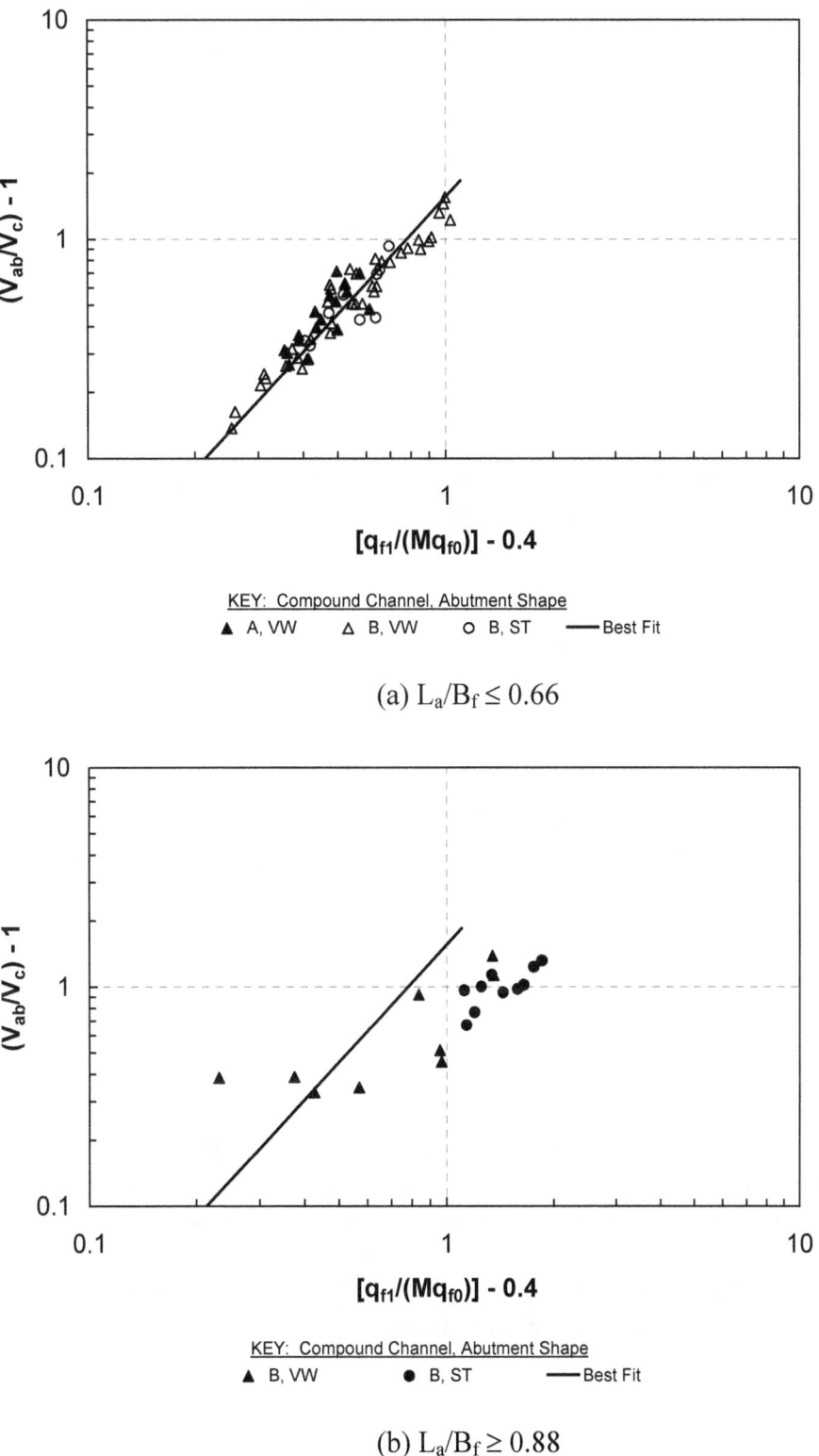

(a) $L_a/B_f \leq 0.66$

(b) $L_a/B_f \geq 0.88$

Figure 30. Relationship between local and approach hydraulic variables.

find in practice if there is significant upstream backwater, which leads to a rise in stage and a reduction in approach velocity below the value required to carry sediment.

The live-bed scour analysis is based initially on contraction scour. The basic condition that must be satisfied for equilibrium live-bed scour comes from a continuity of sediment transport rates between the approach main channel and the main channel in the contracted section, which may, in general, have different widths. If we use the Meyer-Peter and Mueller bedload function,[75] for example, then it must be true that:

$$q^*_{bv1} B_{m1} = q^*_{bv2} B_{m2} \tag{44}$$

where subscripts 1 and 2 refer to the approach and contracted main channel, respectively, and q^*_{bv} is the dimensionless volumetric sediment transport rate given as:

$$q^*_{bv} = \frac{q_{bv}}{\sqrt{(SG - 1) g d_{50}^3}} = 8 (\tau_* - \tau_{*c})^{3/2} \tag{45}$$

where q_{bv} = volumetric sediment transport rate per unit width, SG = specific gravity of the sediment, d_{50} = median sediment grain size, τ_* = Shields' parameter, and τ_{*c} = critical value of Shields' parameter. Substituting the sediment transport relationship given by equation 45 into equation 44 and rearranging it, we have:

$$\frac{\tau_{*2}}{\tau_{*1}} = (\frac{B_{m1}}{B_{m2}})^{2/3} (1 - \frac{\tau_{*c}}{\tau_{*1}}) + \frac{\tau_{*c}}{\tau_{*1}} \tag{46}$$

Note that for $B_{m2} = B_{m1}$, $\tau_{*2} = \tau_{*1}$ regardless of the value of τ_{*c}/τ_{*1}, although it must be less than or equal to 1 for live-bed scour. If τ_{*c}/τ_{*1} is exactly equal to 1, which must be true at the inception of live-bed scour (threshold live-bed scour), we must again have $\tau_{*2} = \tau_{*1}$.

Now Shields' parameter can be expressed in terms of the shear stress formula and Manning's equation for uniform flow in a wide channel in SI units as:

$$\tau_* = \frac{\tau_0}{(\gamma_s - \gamma) d_{50}} = \frac{n^2 V^2}{(SG - 1) d_{50} y^{1/3}} \tag{47}$$

Then, taking the ratio of Shields' parameter between the approach and contracted sections, and replacing the velocity, V, with Q, using continuity for a rectangular main channel, the result is:

$$\frac{\tau_{*1}}{\tau_{*2}} = (\frac{Q_{m1}}{Q})^2 (\frac{y_{m2}}{y_{m1}})^{7/3} (\frac{B_{m2}}{B_{m1}})^2 \tag{48}$$

where it has been assumed that Manning's n remains constant from the approach to the contracted section in the main channel. Solving for the depth ratio, we have:

$$\frac{y_{m2}}{y_{m1}} = [\frac{\tau_{*1}}{\tau_{*2}}]^{3/7} [\frac{B_{m1}/B_{m2}}{Q_{m1}/Q}]^{6/7} \tag{49}$$

Now, equations 46 and 49 can be solved simultaneously for the scour depth y_{m2} relative to the water surface. The solution is shown graphically in figure 31 in the form of a live-bed scour coefficient, C_{LB}, defined as:

$$\frac{y_{m2}}{y_{m1}} = C_{LB}[\frac{Q}{Q_{m1}}]^{6/7} \tag{50}$$

where

$$C_{LB} = \frac{(\frac{B_{m1}}{B_{m2}})^{6/7}}{[(\frac{B_{m1}}{B_{m2}})^{2/3} (1 - \frac{\tau_{*c}}{\tau_{*1}}) + \frac{\tau_{*c}}{\tau_{*1}}]^{3/7}} \tag{51}$$

From equation 51 and figure 31, it is apparent that the coefficient C_{LB} depends on the ratio of approach shear stress to critical shear stress in the main-channel τ_{*1}/τ_{*c}, and on the ratio of approach to main-channel widths B_{m1}/B_{m2}. As τ_{*1}/τ_{*c} becomes large, the curves approach the horizontal and the coefficient for live-bed contraction scour depth depends only on the main-channel width ratio to the 4/7 power, which is close to the power of 0.59 given in HEC-18. On the other hand, the effect of τ_{*1}/τ_{*c} has to be included when it is small, as pointed out by Froehlich,[88] although he used a different sediment transport relationship than employed herein. If the main-channel width is constant, the ratio τ_{*1}/τ_{*c} has no influence and the coefficient C_{LB} has a value of 1.

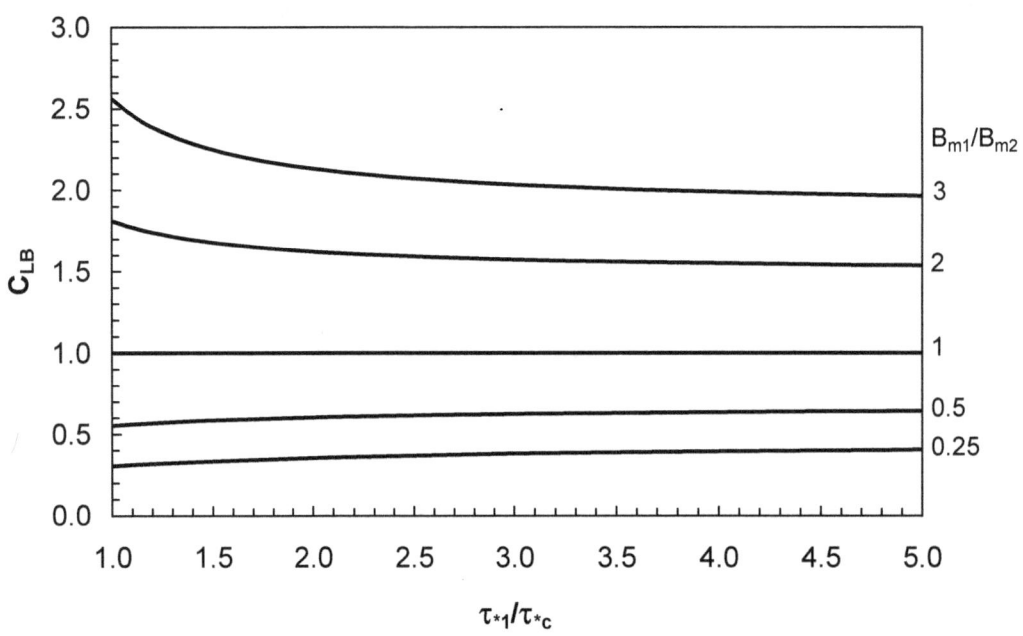

Figure 31. Live-bed contraction scour coefficient for $L_a/B_f = 1.0$.

It is useful at this point to compare the theoretical clear-water contraction scour given by equation 33 and the live-bed contraction scour given by equations 50 and 51. Both equations are for bankline abutments for which the abutment is at the edge of the main channel ($L_a = B_f$). In the limiting case where there is no backwater ($y_{m0} = y_{m1}$) and the approach and contracted main-channel sections have equal widths ($B_{m1} = B_{m2}$), the threshold live-bed scour occurs for $V_{m1} = V_{m0c}$ in equation 33. Under these conditions, equation 33 becomes:

$$\frac{y_{m2}}{y_{m1}} = [\frac{1}{M}]^{6/7} \tag{52}$$

Upon comparison with the live-bed scour equation given by equations 50 and 51, it is clear that equations 50–52 are identical for $C_{LB} = 1$, which is true for $B_{m1} = B_{m2}$ and from the definition of $M = Q_{m1}/Q$. Furthermore, the live-bed scour coefficient remains equal to 1 as τ_{*1}/τ_{*c} increases. Thus, for these conditions, and assuming that local abutment scour is proportional to the theoretical contraction scour, it seems reasonable to estimate live-bed scour for bankline abutments as the threshold of live-bed scour from the clear-water scour equation given by equation 35, in which $C_r' = 2.0$ and $C_0' = 0.47$. In other words, if V_{m1} exceeds V_{m0c}, then V_{m1} is set equal to V_{m0c} in equation 35 for bankline abutments with $B_{m1} = B_{m2}$. By the same reasoning, equation 36 with $V_{f1} = V_{f1c}$ could be used for the unlikely case of live-bed scour around setback abutments in the floodplain. This is the approach that is recommended in this report.

For live-bed scour in the main channel and B_{m1} is not equal to B_{m2}, the data in this report are not directly applicable. However, it has been shown theoretically that live-bed and clear-water contraction scour equations are identical for $B_{m1} = B_{m2}$ and $C_{LB} = 1.0$ at the threshold of live-bed scour. Furthermore, the experimental data taken in this study resulted in equation 35, which predicts a combined abutment and contraction scour of approximately twice ($C_r' = 2$) the theoretical contraction scour for the abutment at or near the bank of the main channel. This suggests that an equation like equation 35 could be combined with the theoretical live-bed coefficient of equation 51 to predict total contraction and abutment live-bed scour for bankline abutments even if B_{m1} is not equal to B_{m2}. Further experiments are needed to confirm this hypothesis.

This theoretical analysis does not account for the observation that a second live-bed scour peak different from the threshold live-bed scour peak ($\tau_{*1}/\tau_{*c} = 1$) can occur for bridge piers at a relative shear stress ratio $\tau_{*1}/\tau_{*c} > 1$.[24,89] The second live-bed scour peak occurs at the bedform transition from dunes to flat bed. Melville[24] showed that for short bridge abutments, the second live-bed scour peak is less than or equal to the threshold live-bed scour peak for uniform sediments, except in the case of ripple-forming sediments ($d_{50} < 0.6$ mm). For nonuniform sediments that armor the streambed by larger particles as the finer particles are washed out of the top bed zones, the threshold live-bed scour peak is less than the second live-bed scour peak, which is itself smaller than the threshold peak for uniform sediments. Whether these conclusions hold for bridge abutments at the edge of the main channel in a compound-channel cross section requires additional experiments.

WSPRO PREDICTIONS

To determine the adequacy of one-dimensional hydraulic computations to be used in formulation I scour predictions, WSPRO simulations of the water-surface profiles and approach velocity distributions for both compound channels A and B are compared to measured laboratory results in this section. Some comparisons are also made with the results from Biglari's two-dimensional numerical model. [54, 55] More extensive comparisons and discussion can be found in a previous report by the author[78] and in a paper by Sturm and Chrisochoides.[86]

In the WSPRO runs for compound channel A, Manning's n was assumed to be constant with depth, with values of 0.020 for the main channel and 0.0174 for the floodplain based on the experimental data for uniform flow. Computed and measured water-surface profiles for compound channel A are compared in figure 32 for a discharge of 0.057 m^3/s and for a vertical-wall abutment having a relative length of $L_a/B_f = 0.33$. The two-dimensional numerical model tracks the experimental results for depth very closely, both upstream and downstream of the abutment, including the expansion zone immediately downstream of the abutment where WSPRO merely produces a constant depth. This is clearly a limitation of the one-dimensional model. Of more interest, however, are the comparisons upstream of the abutment where WSPRO consistently overestimates floodplain depths by about 10 percent.

Predicted velocity distributions in the bridge approach section are compared with measured values in figure 33 for the same experimental conditions as in figure 32. Both WSPRO and the two-dimensional numerical model predict approach floodplain velocities reasonably well. The two-dimensional model seems to capture well the velocity distribution in the interaction zone between the main channel and the floodplain, which was the reason for choosing a k-ε turbulence model with variable turbulent eddy viscosity v_t. However, there are some difficulties with the WSPRO results at the main channel/floodplain interface and at the main-channel centerline. The WSPRO results are computed from the cross-sectional areas of 20 streamtubes having equal conveyances and assumed equal discharges. The vertical wall of the artificial main channel where the depth changes abruptly is apparently the reason for the very low velocity computed by WSPRO near the interface. The WSPRO data points are very close together near the centerline of the main channel because of the large conveyances there.

Comparisons between measured and computed velocity distributions in the bridge contraction section are shown in figure 34. In this section, WSPRO cannot predict the velocity distribution because of the accelerating nonuniform flow in the vicinity of the abutment face. The numerical model, however, shows computed *resultant* velocities at the upstream face of the bridge opening that agree well with the measured resultant velocities.

Computed water-surface profiles for compound channel B are compared with measurements in figure 35 for a relative abutment length of $L_a/B_f = 0.44$ and a discharge of Q = 0.085 m^3/s. In spite of the change in cross-sectional shape and channel roughnesses, the WSPRO comparisons are very similar to those for compound channel A. Downstream of the abutment, the WSPRO depth is nearly constant for uniform flow, while upstream of the abutment, the computed WSPRO depths are larger than the measured values, resulting in a relative error in floodplain depth of about 10 percent. In general, the computed WSPRO depths at the downstream face of

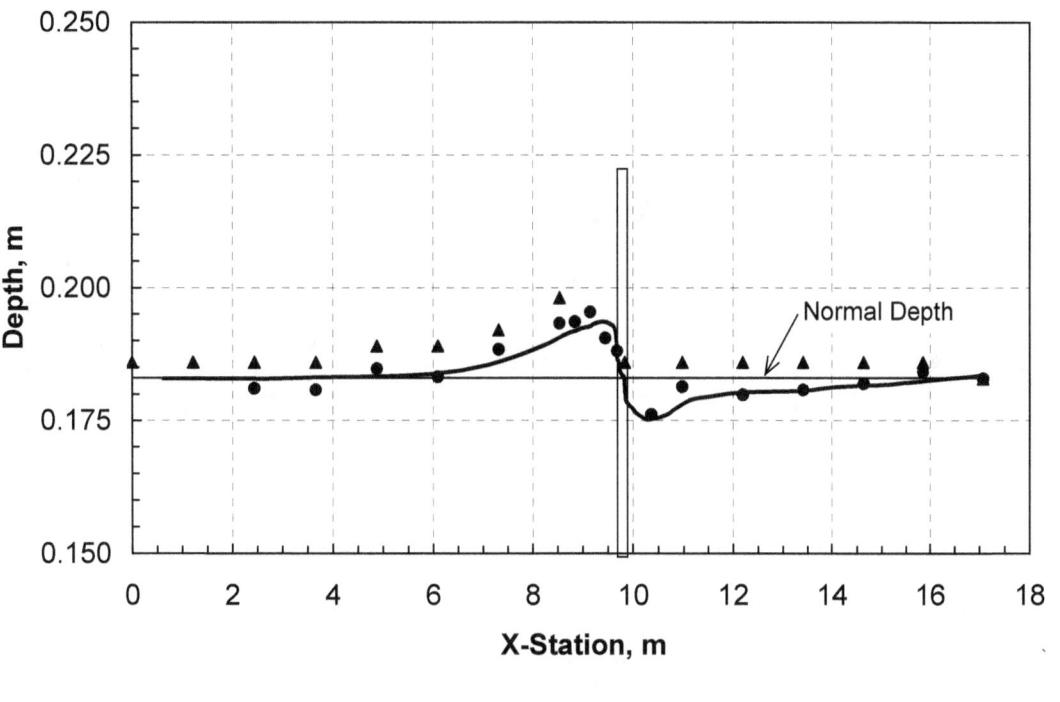

Figure 32. Calculated and measured water-surface profiles for compound channel A. (VW abutment, L_a/B_f = 0.33, Q = 0.057 m^3/s).

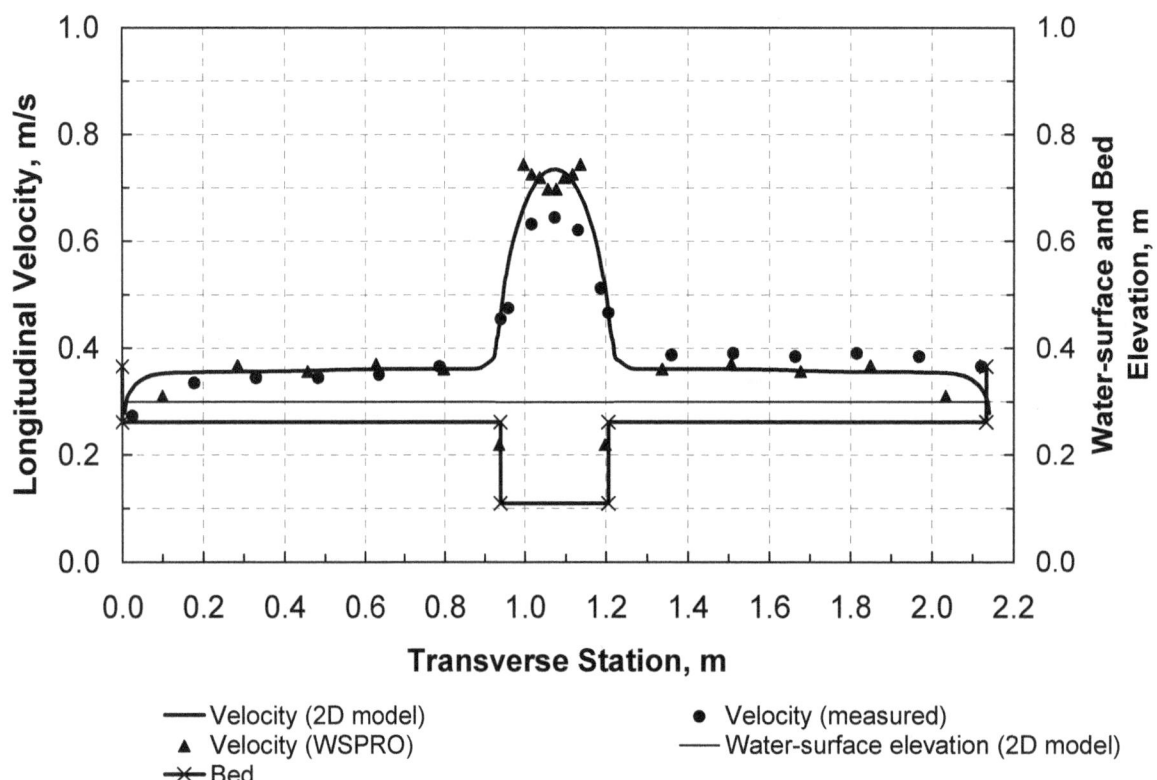

Figure 33. Calculated and measured approach velocity distributions for compound channel A. (VW abutment, $L_a/B_f = 0.33$, Q = 0.057 m^3/s).

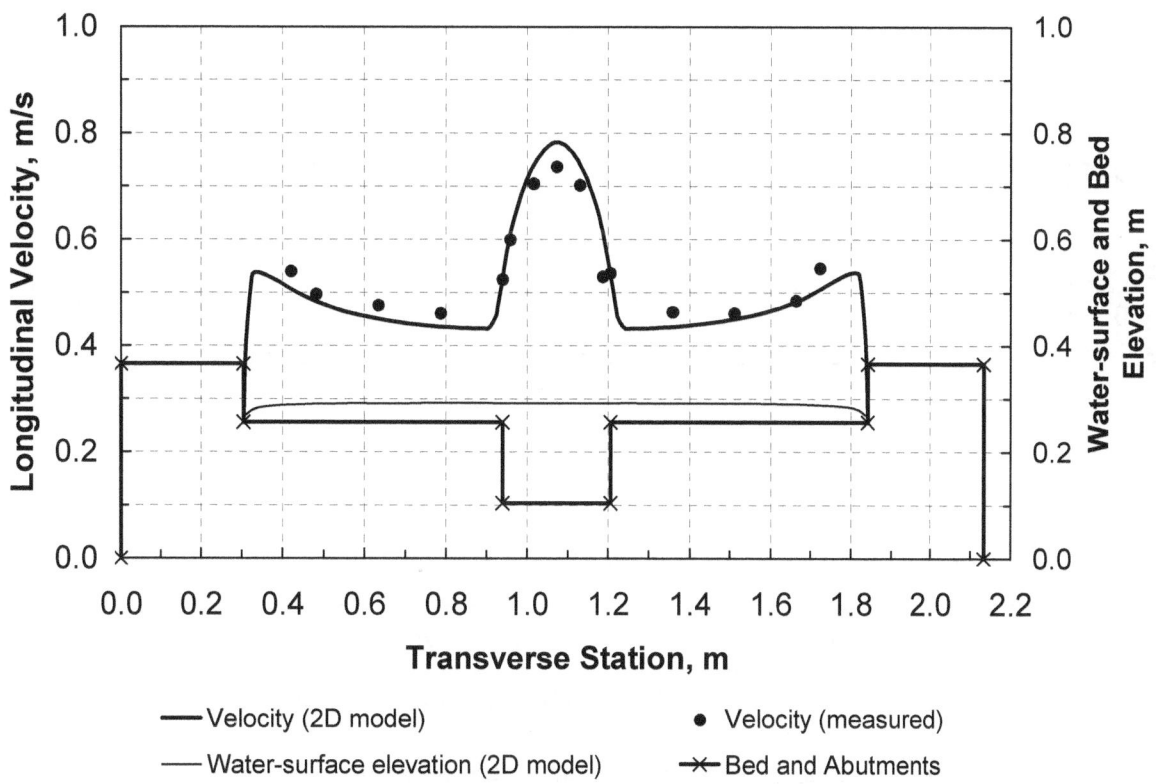

Figure 34. Calculated and measured resultant velocity distributions in the contracted section for compound channel A. (VW abutment, $L_a/B_f = 0.33$, $Q = 0.057$ m^3/s).

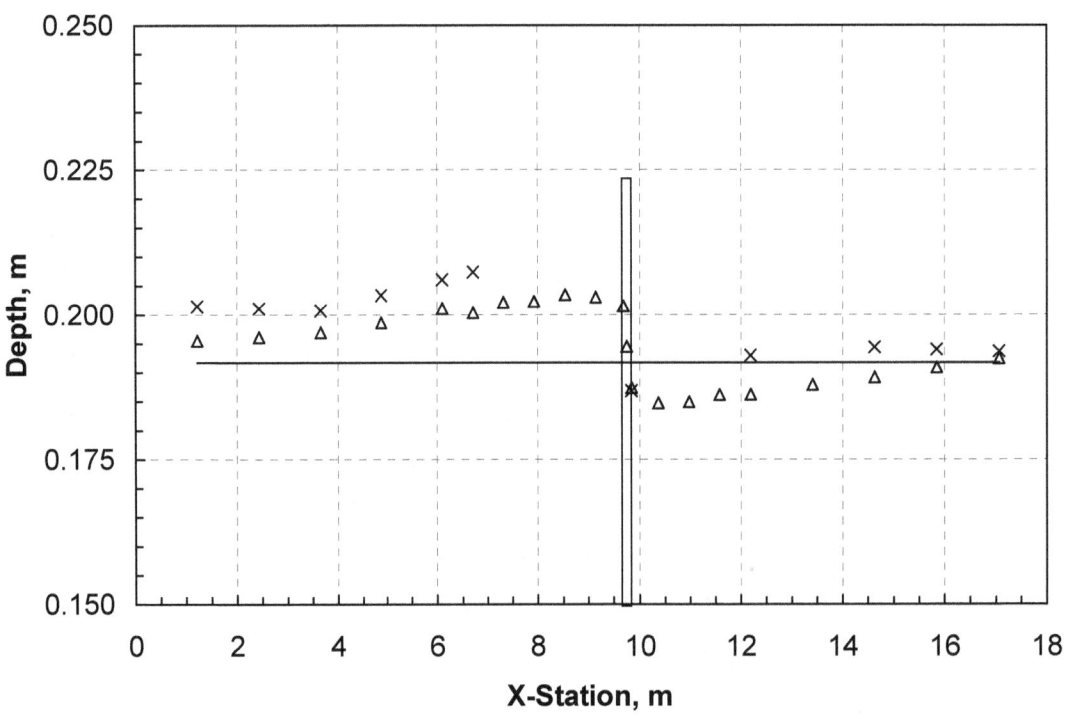

△ Measured Depth —— Normal Depth × Calculated by WSPRO

Figure 35. Calculated and measured water-surface profiles for compound channel B. (VW abutment, $L_a/B_f = 0.44$, Q = 0.085 m^3/s).

the bridge agree rather well with the measured values. Computed and measured approach velocity distributions for compound channel B are shown in figure 36 for the same experimental conditions as in figure 35. There is good agreement between measured and computed WSPRO velocities in the floodplain. However, computed velocities near the interface between the floodplain and the main channel are underpredicted by WSPRO because of the limitations of a one-dimensional model in this region.

More detailed comparisons of computed and predicted values of V_{fl}, the approach floodplain velocity upstream of the end of the abutment; y_{fl}, the approach floodplain depth; and M, the discharge distribution factor, are given in figures 37, 38, and 39 for both cross sections and all discharges with $L_a/B_f \leq 0.88$. The results for both vertical-wall and spill-through abutment shapes are included in these figures. The approach velocity V_{fl} is predicted well by WSPRO for compound channel B as shown in figure 37, with a root mean square deviation between the measured and calculated values of about 0.015 m/s. The WSPRO comparisons for V_{fl} in compound channel A are somewhat scattered; however, the two-dimensional numerical turbulence model performs reasonably well for compound channel A. The approach depth y_{fl} is consistently overestimated by WSPRO in both compound channels A and B as shown in figure 38. The mean percent error is about 12 percent. The two-dimensional numerical model slightly underpredicts the approach floodplain depths.

In figure 39, computed and predicted values are shown for the discharge distribution factor M, which is defined as the ratio of the discharge, in that portion of the approach cross section having a width equal to the bridge opening width, to the total discharge. Thus, the prediction of M depends on the prediction of the approach depth and velocity distributions. It can be observed that M is predicted equally well by WSPRO and the two-dimensional numerical model in compound channel A; however, WSPRO underestimates M for the smaller values in compound channel B. This problem is caused by the differences in the velocity distributions noted earlier for compound channel B, as well as in the overprediction of y_{fl} by WSPRO. The results for M show that at least some of the effect of the main channel/floodplain interaction has been accounted for by using measured Manning's n values; however, this achieves only the correct split between the main-channel and floodplain flow, not necessarily the correct detailed velocity distribution near the interface. This is a problem with the one-dimensional analysis afforded by WSPRO that can be overcome by a two-dimensional numerical model.

As a final comparison, the calculated and measured velocities in the bridge contraction section near the abutment face are shown in figure 40 for compound channel B. The measured velocities V_{ab} are the measured maximum resultant velocities found near the abutment face. The velocities calculated by WSPRO for this case necessarily come from the bridge section at the downstream face of the bridge opening and are taken as the mean velocity in the streamtube nearest the abutment face. WSPRO clearly cannot predict velocities in this part of the accelerating flow field and would not be expected to do so. *Based on these results, it is inadvisable to use WSPRO to predict velocities near the abutment face for use in scour-prediction equations.*

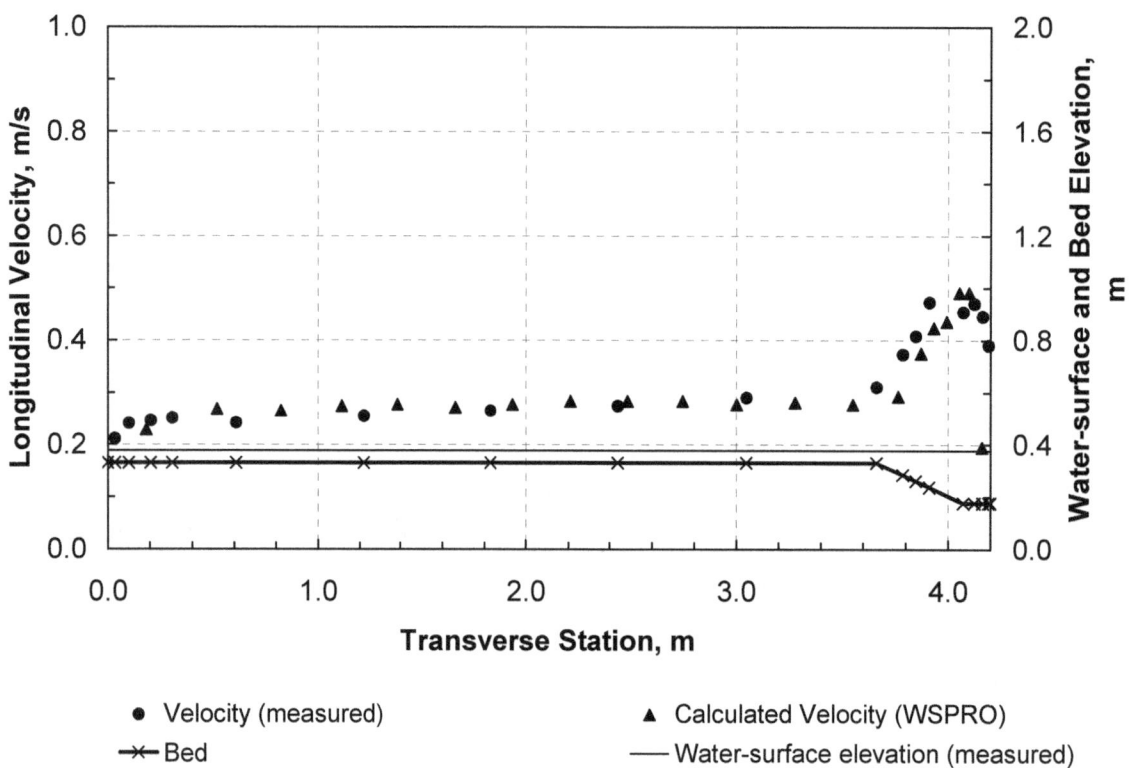

Figure 36. Calculated and measured approach velocity distributions for compound channel B. (VW abutment, $L_a/B_f = 0.44$, Q = 0.085 m^3/s).

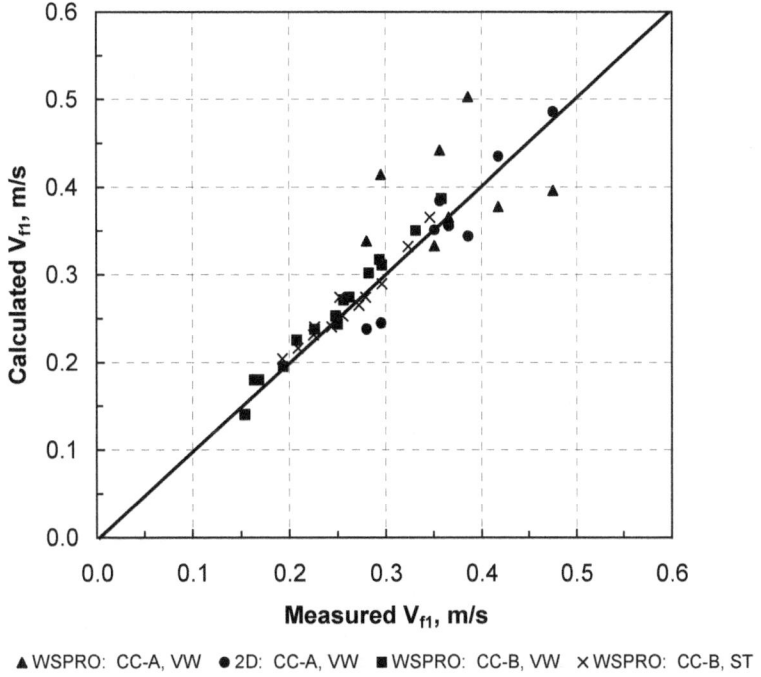

Figure 37. Comparisons of approach floodplain velocities (CC = compound channel).

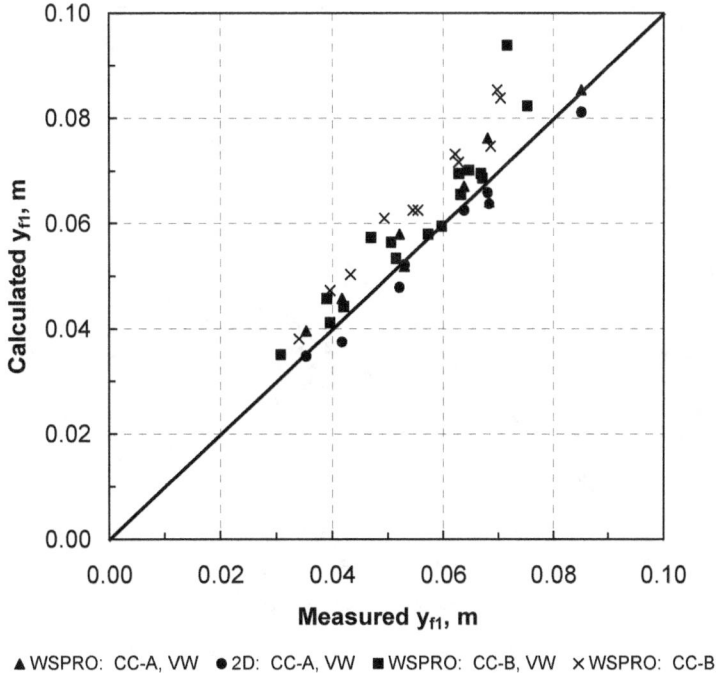

Figure 38. Comparisons of approach floodplain depths (CC = compound channel).

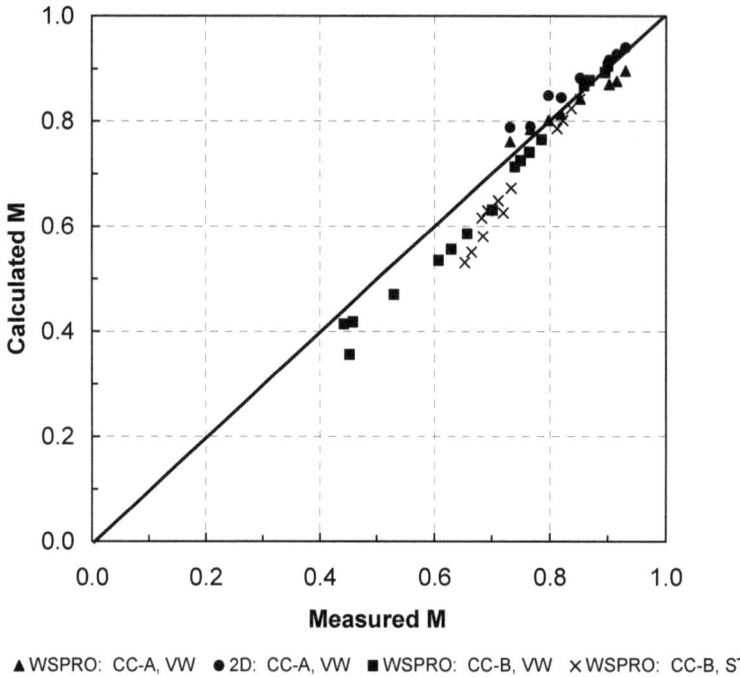

Figure 39. Comparisons of discharge distribution factor M (CC = compound channel).

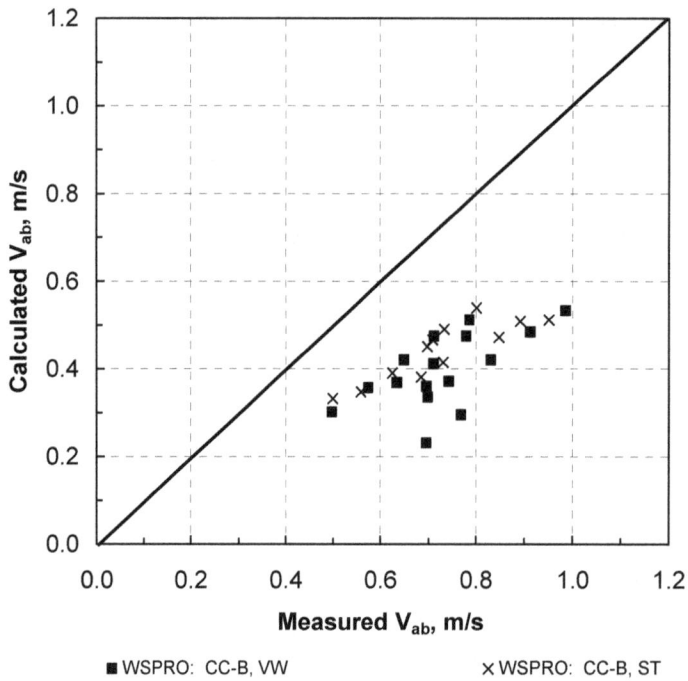

Figure 40. Comparisons of maximum velocity at abutment face
(CC = compound channel).

TIME DEVELOPMENT RELATIONSHIP

Based on the dimensional analysis result given by equation 45, time can be added as an independent variable in the scour development process to produce an expected relationship of the form:

$$\frac{d_{st}}{y_{f0}} = f\left[\frac{V_{ab}}{V_c}, \frac{V_c\, t}{y_{f0}}\right] \tag{53}$$

where d_{st} = scour depth at any time t. The influence of y_{ab}/y_{f0} has not been included in equation 48 based on the experimental results. The experimental measurements of scour depth with time are presented according to equation 53 for three sediment sizes (d50 = 3.3 mm, 2.7 mm and 1.1 mm) in figure 41. These results for the time development of scour include only those cases for which the maximum scour depth developed near the upstream corner of the abutment as discussed previously in chapter 3. However, the location of the scour hole at the upstream corner occurred for the larger discharges so that the worst cases are included in these results. The curves for different values of V_{ab}/V_c have a functional form that begins with a linear development of scour depth with the logarithm of time, followed by an abrupt leveling off to a nearly constant value equal to the equilibrium scour depth that depends only on V_{ab}/V_c.

The functional behavior and collapse with respect to sediment size shown in figure 41 suggest the possibility of a universal set of time-development curves that can be applied to field cases provided that the dimensionless variables fall in the same range as in the laboratory experiments. Accordingly, a least-squares regression analysis was applied to the data and suggested interpolated curves were developed and plotted as solid lines in figure 41. Thus, for a given sediment size, which determines V_c; a given abutment velocity V_{ab}, which is determined by the abutment shape, degree of floodplain contraction, and flow velocity distribution; and a given time corresponding to the design flood duration, an estimated depth of scour can be obtained. For example, if $V_{ab}/V_c = 1.5$, approximately two-thirds of the equilibrium scour depth is reached in a flow duration of only about 10 percent of the equilibrium time. Figure 41 shows very clearly the interplay of time, flow distribution, and sediment size in determining a design value of abutment scour depth.

COMPARISON OF SCOUR FORMULAS WITH MEASURED DATA

Several formulas for clear-water abutment scour depth, in addition to the one formulated herein, have been proposed in the literature. These were discussed in detail in chapter 2 and the formulas will be referred to as: (1) Melville,[14,24] (2) Froehlich clear-water scour (CWS),[12] (3) Froehlich live-bed scour (LBS),[12] (4) GKY,[66] and (5) Maryland.[68,70] The formulas proposed by Melville and Froehlich both include data collected in Auckland, New Zealand, in rectangular flumes. The GKY formula relies on data from Lim[67] and Sturm and Janjua,[20] while the Maryland procedure is based on experimental data in rectangular flumes assembled by Palaviccini.[69] Because the scour data collected in this study include both vertical-wall and spill-through abutments, a shape factor was applied for each formula. The shape factor for vertical-wall abutments was taken to be 1.0 for all formulas. The shape factor K_{ST} for spill-through abutments in the Melville formula

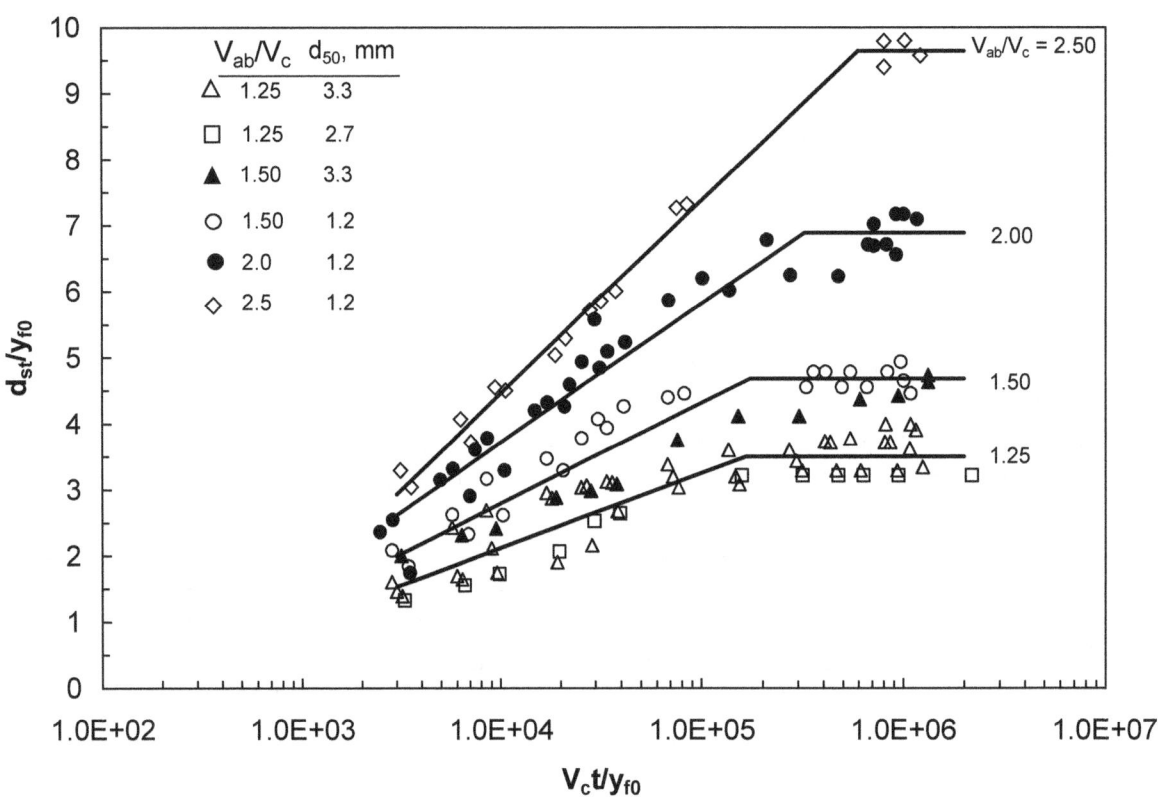

Figure 41. Dimensionless representation of time development of scour.

varies linearly from 0.45 (for 1.5:1 sideslopes) to 1.0 as L_a/y_1 varies from 10 to 25 and is constant outside of these limits with $K_{ST} = 0.45$ for $L_a/y_1 \leq 10$ and $K_{ST} = 1.0$ for $L_a/y_1 \geq 25$. (Melville does not give a shape factor for 2:1 sideslopes, so the closest value of 1.5:1 was used in the predictions to follow.) For the other formulas, a standard constant shape factor of 0.55 for spill-through abutments was used based on the guidance given by HEC-18. The formula from the present study (equation 36 with $C_r = 8.14$ and $C_0 = 0.40$) is also compared with the experimental data using the shape factor for spill-through abutments derived herein as equation 37.

The comparisons between the measured and predicted scour depths for the five formulas listed above are given in figures 42 through 46, for a total of 74 data points measured in the present study in a compound channel (compound channel B). The line of perfect agreement is shown as a solid line, and the two dashed lines indicate a variation of ±30 percent from the line of perfect agreement in order to gauge the degree of scatter in the comparisons.

The Melville formula overpredicts scour depths in the lower range of the data by several hundred percent as shown in figure 42. Only about one-half of the data points fall within the ±30 percent range of variation relative to perfect agreement. The Froehlich CWS formula shown in figure 43 shows a similar pattern to the Melville formula, which is to be expected because they are based on a similar data set. However, several larger values of scour are underpredicted by the Froehlich CWS formula because of the constant spill-through abutment shape factor of 0.55, whereas the Melville formula has a shape factor of 1.0 for the larger values of L_a/y_{f1}. The Froehlich LBS formula, which is recommended by HEC-18, even for clear-water scour, is given in figure 44. In general, it shows similar data scatter in comparison to the Froehlich CWS formula, but with greater overprediction of nearly all scour depths. The GKY and Maryland formulas in figures 45 and 46, respectively, display a trend opposite to that of the first three formulas. They both tend to underpredict the measured scour depth by considerably more than the lower 30-percent range. In order to maximize the predicted values for both formulas, the entire approach floodplain flow was assumed to pass through the contracted-section floodplain for $L_a/B_f \leq 0.88$, while the total flow was assumed to pass through the main channel in the contracted section for $L_a/B_f = 0.97$ and 1.0. In general, the Maryland formula predicts greater scour depths than the GKY formula, and both formulas suffer from the use of the spill-through shape factor of 0.55 when the present study as well as the experiments by Melville[24] show that it approaches 1.0 as the relative abutment length, or the fraction of blocked flow, becomes large. The GKY formula was applied in SI units with the leading constant factor of 1.37 as recommended in reference 66; however, a value approximately twice as large would greatly improve the performance of the formula. Finally, the formula recommended in this study is compared with the experimental data in figure 47. While the formula itself is based on the data used in the comparison, figure 47 is indicative of the overall performance of the formula. Only 6 of the 74 data points lie completely outside the ±30 percent range. Four of these outliers occur for measured scour depths on the order of 5 cm or less, which are more difficult to measure and reproduce because of the inherent uncertainty in the experiments. In conclusion, it seems that the formulas based primarily on experiments in rectangular flumes cannot reproduce the physics of the flow and the coupled flow distribution and scour behavior associated with the compound channel.

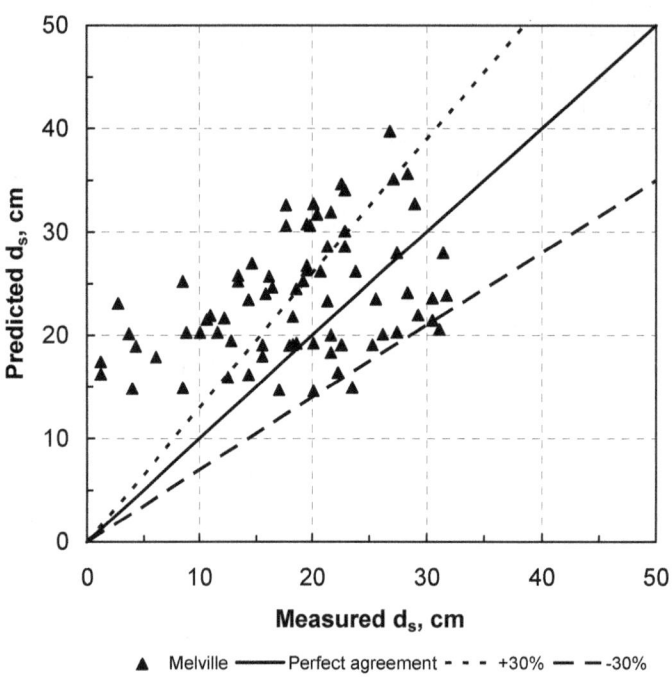

Figure 42. Comparison of measured and predicted scour depths using the Melville formula.

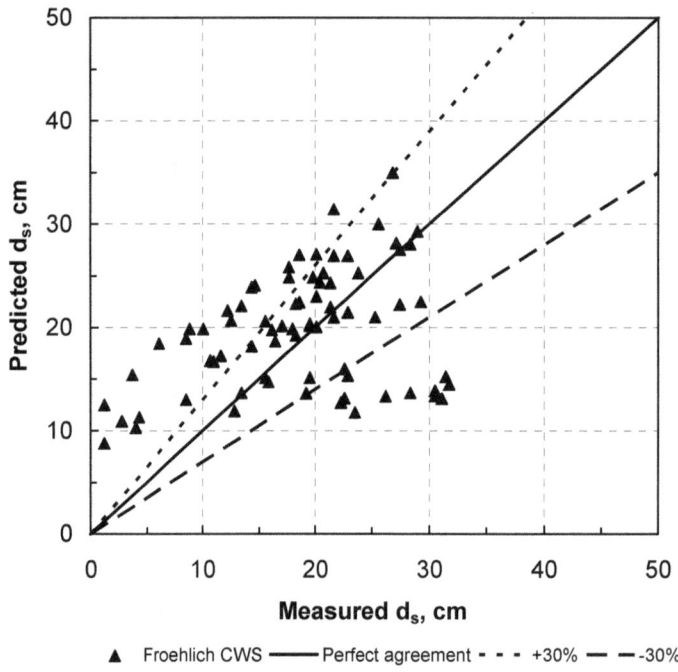

Figure 43. Comparison of measured and predicted scour depths using the Froehlich clear-water scour formula.

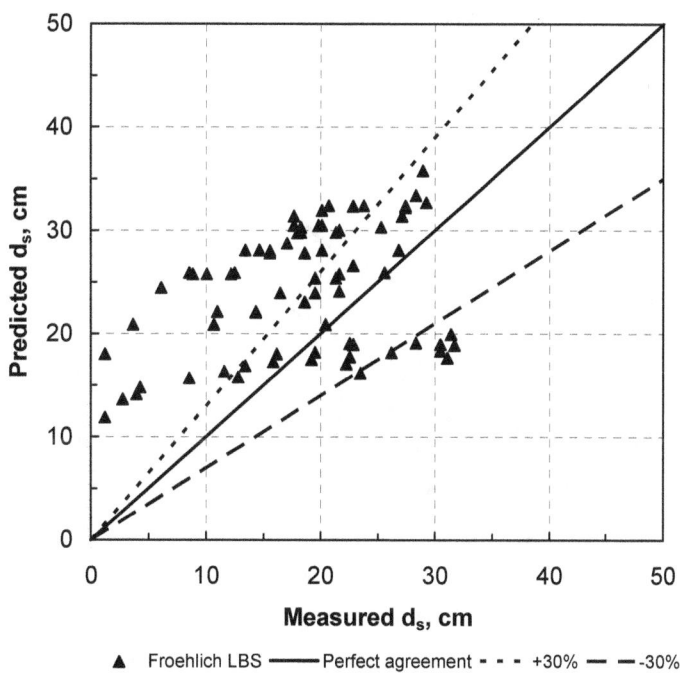

Figure 44. Comparison of measured and predicted scour depths
using the Froehlich live-bed scour formula.

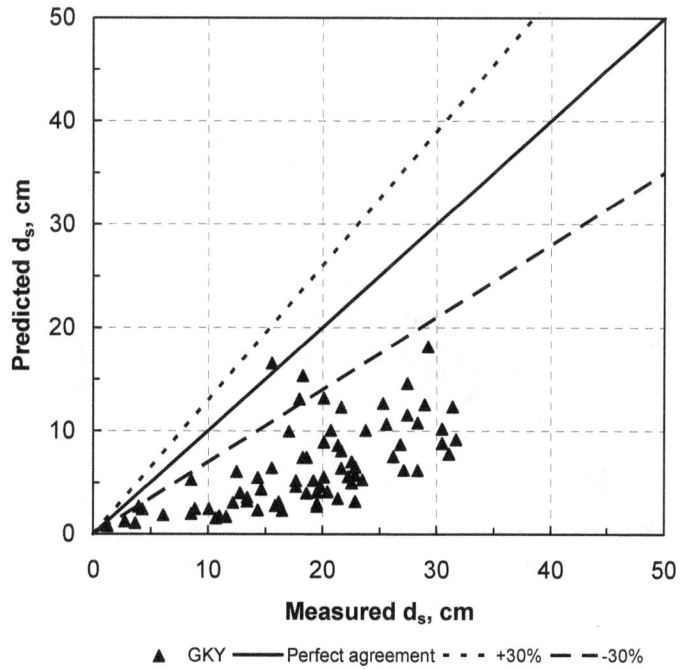

Figure 45. Comparison of measured and predicted scour depths
using the G. K. Young (GKY) formula.

91

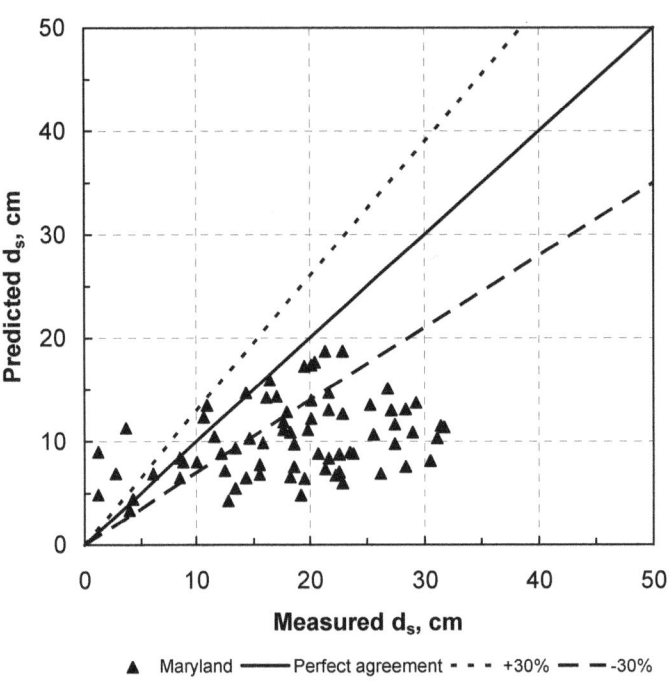

Figure 46. Comparison of measured and predicted scour depths
using the Maryland formula.

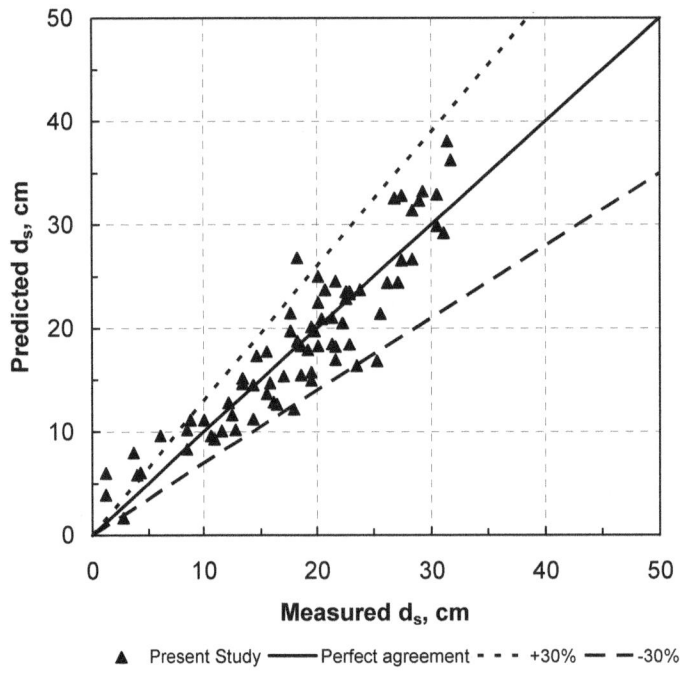

Figure 47 . Comparison of measured and predicted scour depths
using the formula from the present study.

PROPOSED IMPLEMENTATION PROCEDURE FOR ABUTMENT SCOUR PREDICTION

As a result of the experimental results on the equilibrium depth of clear-water abutment scour given in figure 27 and on the time development of scour shown in figure 41, a prediction procedure that accounts for discharge distribution, sediment size, and time can be developed. Based on the extensive comparisons between WSPRO estimations of scour parameters and parameter estimation by a two-dimensional numerical turbulence model, it is assumed that the WSPRO predictions provide a reasonable estimation of the approach-flow hydraulic parameters for the prediction of clear-water abutment scour. It is further assumed that the abutment velocities to be used in the scour time development determination in figure 41 can be obtained from the relationship developed in figure 30, which depends only on parameters calculated from the WSPRO results. First, the formulas recommended in the scour-prediction procedure are repeated below. Bridge abutments that terminate at the edge of the main channel are referred to as bankline abutments, while abutments located on the floodplain are called setback abutments.

1.   Definition of discharge contraction ratio, M (equation 19 from chapter 3):

$$M = \frac{Q_{m1} + (Q_{f1} - Q_{obst1})}{Q} \tag{54}$$

where Q_{m1} = discharge in the approach main channel, Q_{f1} = discharge in the approach floodplain, Q_{obst1} = obstructed floodplain discharge over a length equal to the abutment length in the approach cross section, and Q = total discharge = $Q_{m1} + Q_{f1}$. The value of M as defined by equation 54 can be calculated from the WSPRO output as $1 - M(K)$. For road overflow, a consistent definition of M would require the denominator to include only that portion of the total Q going through the bridge opening.

$$\frac{V_c}{\sqrt{(SG - 1)g\, d_{50}}} = 5.75 \sqrt{\tau_{*c}} \, \log\left[\frac{12.2\, y}{2\, d_{50}}\right] \tag{55}$$

2.   Keulegan's equation for critical velocity (equation 22 from chapter 3):

where V_c = critical velocity, SG = specific gravity of sediment, d_{50} = median sediment grain diameter, y = flow depth, and τ_{*c} = critical value of Shields' parameter. If the channel is not very wide, then the depth, y, should be replaced by the hydraulic radius, R. Equation 55 only applies to fully rough turbulent flow. For sediment sizes of less than about 1.2 mm, the Einstein correction factor given on p. 259 of reference 75 should be used to multiply the logarithmic argument.

3.   Scour-prediction formula for clear-water scour around setback and bankline abutments, including shape correction K_{ST} and safety factor FS (modified equation 36 from chapter 4):

$$\frac{d_s}{y_{f0}} = K_{ST} C_r \left(\frac{q_{f1}}{M V_{xc} y_{f0}} - C_0\right) + FS \tag{56}$$

where K_{ST} = spill-through abutment shape factor from equation 57 below; $C_r = 8.14$; $C_0 = 0.40$; $q_{f1} = V_{f1} y_{f1}$; V_{f1} and y_{f1} = average floodplain velocity and depth, respectively, in the obstructed portion of the floodplain in the bridge approach-flow section for constricted flow; $V_{xc} = V_{f0c}$ for abutments located on the floodplain and $V_{xc} = V_{m0c}$ for abutments near the bank of the main channel, where V_{f0c} and V_{m0c} are critical velocities for the unconstricted flow in the floodplain and main channel, evaluated for depths of y_{f0} and y_{m0}, respectively; and y_{f0} and y_{m0} are the unconstricted flow depths in the floodplain and main channel at the approach section. A value of FS = 1.0 is recommended because this is greater than the standard error of 0.75 for d_s/y_{f0} for the best fit of the experimental data. If V_{f1} is greater than the critical velocity V_{f1c} in the approach floodplain flow, then it should be set equal to the critical velocity. Finally, if $q_{f1}/(MV_{xc}y_{f0}) > 1.6$, then d_s/y_{f0} should be set equal to the maximum value of 10. The elevation of the bottom of the scour hole is taken to be a distance of $d_s + y_{f0}$ below the unconstricted water-surface elevation at the bridge.

4.   Spill-through abutment shape factor (equation 37 from chapter 4):

$$K_{ST} = 1.52 \frac{\xi - 0.67}{\xi - 0.40} \quad for \quad 0.67 \le \xi \le 1.2 \tag{57}$$

where $\xi = q_{f1}/(MV_{xc}y_{f0c})$ and $K_{ST} = 1.0$ for $\xi \ge 1.2$ and 0 for $\xi \le 0.67$.

5.   Scour-prediction formula for clear-water and live-bed scour around bankline abutments (equation 35 from chapter 4):

$$\frac{d_s}{y_{f0}} = C_r' \left[\frac{q_{m1}}{M(V_{m0c} y_{f0})} - C_0'\right] + FS \tag{58}$$

where $q_{m1} = V_{m1} y_{m1}$; V_{m1} and y_{m1} = mean velocity and depth, respectively, in the approach main channel; $C_r' = 2.0$; $C_0' = 0.47$; and FS = 1.0. Equation 58 is based on clear-water scour data around bankline abutments for which the abutment shape factor was equal to 1.0. In this report, equation 56 is recommended for clear-water scour around bankline abutments instead of equation 58. However, equation 58 is suggested as an interim equation for live-bed scour around bankline abutments with $V_{m1} = V_{m1c}$ until the data for the live-bed scour case are available.

6. Conversion from approach hydraulic variables to local hydraulic variables for use in figure 41(equation 43 from chapter 4):

$$\frac{V_{ab}}{V_c} - 1 = 1.56 \left(\frac{q_{f1}}{M \, q_{f0c}} - 0.4\right)^{1.78} \tag{59}$$

where $q_{f0c} = V_{xc}y_{f0c}$; V_{ab} = resultant depth-averaged velocity near the upstream corner of the abutment face; and V_c = critical velocity at the same location, which is determined from Keulegan's equation for the flow depth in the bridge at the toe of the abutment. The values of $V_c t/y_{f0}$ and V_{ab}/V_c are required to use figure 41.

The steps in the proposed procedure are:

1. From the field data, obtain at least one surveyed cross section, and preferably three sections (bridge exit, downstream face of the bridge, and bridge approach). Also estimate Manning's n for the floodplains and the main channel, and obtain bridge geometry data and sediment size d_{50}.

2. Determine the 100-year and 500-year design discharges based on drainage area and regional frequency estimates (available from the National Flood Frequency (NFF) program in version 6 of the highway drainage integrated computer system (HYDRAIN) software which was funded by 31 State highway agencies.

3. Run WSPRO and obtain M, V_{f1}, y_{f1}, and y_{f0} from the results. Calculate V_{f1} and y_{f1} as the average values in the obstructed portion of the floodplain in the approach section for the constricted flow. The value of y_{f0} is obtained in a similar manner for the unconstricted flow. Estimate the value of V_{f0c} from y_{f0} and d_{50} for setback abutments, or V_{m0c} from R_{m0} and d_{50} for bankline abutments, using Keulegan's equation (equation 55), where R_{m0} = hydraulic radius of the main channel. Also calculate the value of V_{1c} in the approach flow from Keulegan's equation for either the floodplain or the main channel, depending on whether the abutments are setback or bankline abutments, respectively.

4. Calculate the dimensionless clear-water scour depth d_s/y_{f0} from equation 56, including the abutment shape factor from equation 57 and FS = 1.0. The maximum value of d_s/y_{f0} should not be taken as being any greater than 10 based on the experimental results. In addition, if $V_{f1} \geq V_{f1c}$ for setback abutments, then threshold live-bed scour occurs, so set $V_{f1} = V_{f1c}$. The elevation of the bottom of the scour hole is taken to be a distance of $d_s + y_{f0}$ below the unconstricted water-surface elevation at the bridge.

5. If it is a bankline abutment, check for the possibility of live-bed scour by determining whether $V_{m1} \geq V_{m1c}$. If so, then set $V_{m1} = V_{m1c}$ in equation 58 and solve for the live-bed scour depth. For the bankline abutment, regardless of whether the scour is clear-water or live-bed, the scour depth calculated from the equations recommended herein includes

both contraction and local abutment scour.

6. From the watershed size and a hydrologic estimate of the lag time of the watershed, use NFF in HYDRAIN to generate a hydrograph corresponding to the peak design discharge from which a flood duration can be estimated.

7. For the duration obtained from step 6, determine the percentage of the equilibrium scour depth that will occur from figure 41. To estimate V_{ab}/V_c as needed in figure 41, use equation 59. This time-dependent scour-depth calculation is only intended to temper engineering judgment and should not necessarily be used to reduce the estimated equilibrium scour depth, because scour can be cumulative.

Example

The following example has been adapted and modified from HEC-18, and it was first given in a report by the author.[78] However, it has been converted to SI units and modified to include the spill-through abutment shape factor in comparison to the presentation in that report.

A bridge with a 228.6-m (750-ft) opening length spans Burdell Creek, which has a drainage area of 971 square kilometers (km^2) (375 mi^2). The exit and bridge cross sections are shown in figure 48 with three subsections and their corresponding values of Manning's n. The slope of the stream reach at the bridge site is constant and equal to 0.001 m/m. The bridge has a deck elevation of 6.71 m (22.0 ft) and a bottom chord elevation of 5.49 m (18.0 ft). It is a type 3 bridge with 2:1 abutment and embankment slopes, and it is perpendicular to the flow direction (zero skew). The tops of the left and right spill-through abutments are at X-stations of 281.9 m and 510.5 m, respectively, and the abutments can be considered to be setback abutments. There are six bridge piers, each with a width of 1.52 m (5.0 ft). The sediment has a median grain diameter d_{50} of 2.0 mm (0.00656 ft). The requirement is to estimate the clear-water abutment scour for the 100-year design flood.

Solution

From HYDRAIN, the NFF program can be used to estimate the design flood flows. For the given drainage area and for a given region of the country (coastal plain in Georgia), the predicted Q_{100} = 397 m^3/s (14,000 ft^3/s) and the predicted Q_{500} = 567 m^3/s (20,000 ft^3/s). Calculations are done in this example for Q_{100}.

The WSPRO input data file is shown as table 4 in appendix A for Q_{100} = 397 m^3/s. The program was actually run twice, first to obtain the water-surface elevations for both the unconstricted and the constricted cases at the approach cross section, and second, with the HP 2 data records (hydraulic parameter data record for a particular cross section of the WSPRO output file) to compute the velocity distribution in the approach section for the unconstricted (undisturbed) water-surface elevation of 4.038 m (13.25 ft) and the constricted water-surface elevation of 4.157 m (13.64 ft). These elevations can be extracted from table 5, where the results of the water-surface profile computations are given. Tables 6 and 7 give the results of the velocity distribution computations for the unconstricted and constricted cases, respectively.

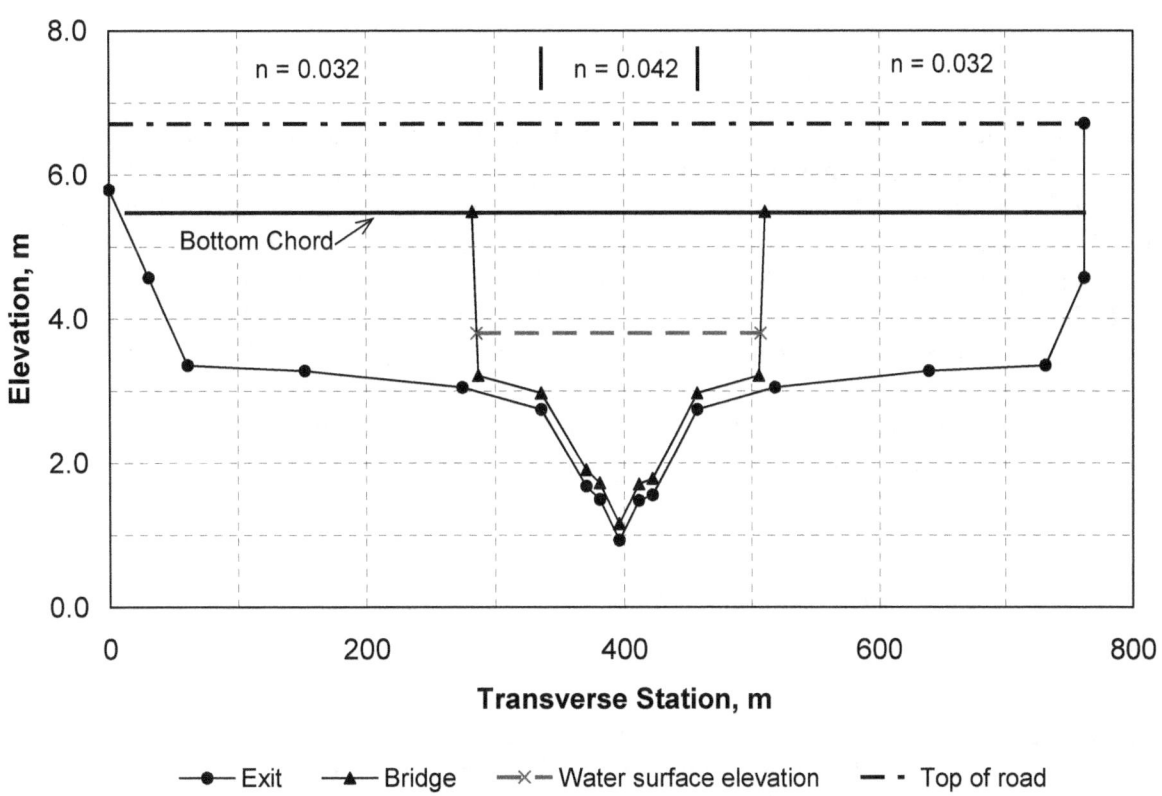

Figure 48. Cross sections and water-surface elevation for Burdell Creek bridge.
(Drainage area = 971 km^2, Q$_{100}$ = 397 m^3/s).

Now the scour parameters can be calculated from the WSPRO results. The value of M(K) from table 5 is 0.189, which by definition gives M = 1 − M(K) = 0.811. For consistency with the current FHWA methodology, the unconstricted and constricted floodplain depths and velocities are determined by the procedure given in HEC-18 for Froehlich's equation. First, for the *constricted* case, the abutment length is determined from table 5 as the difference between the left edge of the water (LEW) for the BRDG and APPR sections:

$$L_a = LEW_{BRDG} - LEW_{APPR} = 232.8\,m\ (764\,ft) \tag{60}$$

Then, from table 7 for the constricted flow, the blocked flow in the approach section up to X STA. = LEW_{BRDG} is 1.97 streamtubes by interpolation. Therefore, the blocked flow Q_{fl} is $(1.97/20)397 = 39.1\ m^3/s\ (1380\ ft^3/s)$, because each of the 20 streamtubes carries 1/20 of the total flow. Also, the blocked flow area A_{fl} can be interpolated from table 7 to be 106.8 m^2 (1150 ft^2). Now we can calculate $V_{fl} = Q_{fl}/A_{fl} = 39.1/106.8 = 0.366$ m/s (1.20 ft/s), and $y_{fl} = A_{fl}/L_a = 106.8/233 = 0.458$ m (1.50 ft). In a similar manner, the value of y_{f0} is found for the equivalent blocked flow area in the *unconstricted* cross section from table 6 to be 0.357 m (1.17 ft).

The critical velocities for coarse sediments are determined by substituting into equation 55 (Keulegan's equation). For the constricted approach section, we have for a floodplain depth of 0.458 m (1.50 ft):

$$V_{f_{1c}} = \sqrt{(2.65-1)(9.81)(0.002)(0.035)} * 5.75 * \log\frac{(12.2)(0.458)}{2*0.002} = 0.61\,m/s \tag{61}$$

where Shields' parameter was taken to be 0.035 for this sediment size.[75] Because $V_{fl} < V_{f1c}$, it is apparent that we have clear-water scour. In a similar manner, the value of V_{f0c} for an unconstricted floodplain depth of 0.357 m (1.17 ft) is 0.59 m/s (1.93 ft/s).

To compute the scour depth for the setback abutments, substitute into equation 56 to obtain:

$$\frac{d_s}{y_{f0}} = 8.14\,K_{ST}\,[\frac{(0.366)(0.457)}{(0.81)(0.61)(0.357)} - 0.4] + 1.0 = 4.44 \tag{62}$$

where shape factor $K_{ST} = 0.77$ from equation 57. Finally, the scour depth is 4.44 x 0.357 = 1.59 m (5.22 ft). In general, this calculation would be repeated for the right abutment; however, this example has an essentially symmetrical cross section.

The NFF program can be used to develop a design flood hydrograph. Assuming that the watershed is 48.3 km (30 mi) long with a lag time of 15 hours, the resulting hydrograph can be computed and is shown in figure 49. As a conservative assumption, the flood duration is estimated as the time required for a constant discharge equal to the peak discharge to give the

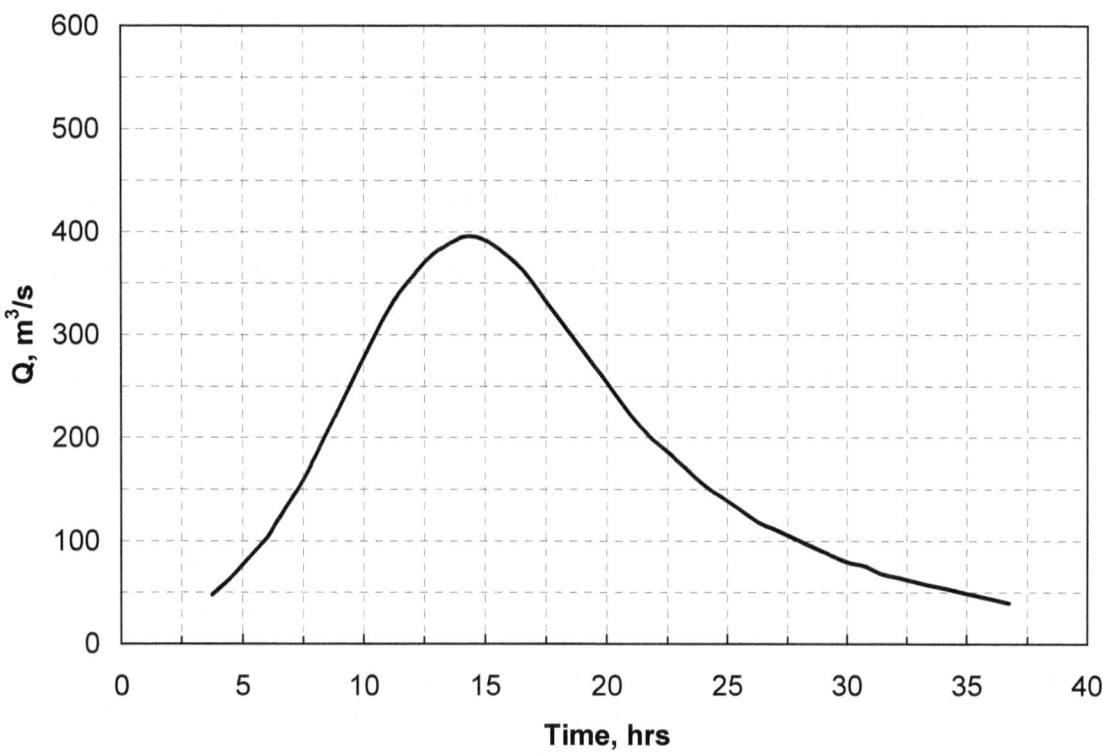

Figure 49. Design 100-year hydrograph for Burdell Creek.
(Drainage area = 971 km^2, lag time = 15 hours).

same volume of direct runoff as the original hydrograph. This results in a duration of 13.4 hours.

The critical velocity at the toe of the abutment is computed from a depth equal to the water-surface elevation in the bridge opening, which is 3.80 m (12.47 ft) from table 5 minus the ground elevation at the toe (3.15 m) to give y_{ab} = 0.646 m (2.12 ft). Then, using equation 55 again for consistency, the critical velocity V_c = 0.64 m/s (2.1 ft/s), and $V_c t/y_{f0}$ = (0.64)(13.4)(3600)/(0.357) = 9 x 10^4. From equation 59, V_{ab}/V_c = 1.54, and figure 41 indicates that approximately 90 percent of the equilibrium scour will occur over the estimated flood duration of 13.4 hours. Under these circumstances, the reduction in scour caused by equilibrium not being reached is small so that the final result for the clear-water abutment scour depth is left at the value of 1.59 m (5.22 ft) that was previously estimated.

Because scour holes can develop as a result of several floods over time, the time analysis proposed herein is intended as a judgment factor in reducing the estimated scour only when: (1) the estimated scour is significantly higher than that actually observed on an existing bridge, or (2) the watershed is so small that only a small fraction of equilibrium scour is reached in a typical design flood. It is strongly recommended that a whole range of discharges be considered and that the percentage of equilibrium scour be investigated in each case. Figure 41 provides the first method for estimating the percentage of equilibrium scour for floods of differing magnitude, and it should greatly enhance the engineer's judgment in making a final evaluation of scour susceptibility.

It should be noted that the estimated equilibrium scour has been determined for a spill-through abutment with a shape factor of 0.77 for this particular example, which is greater than the value of 0.55 currently recommended by HEC-18. However, the shape factor varies and can be as large as 1.0 for long spill-through abutments as indicated by equation 57 (shape has no influence in some cases). In addition, the estimated scour could be increased by a skewness factor for abutments not perpendicular to the flow as recommended by HEC-18; however, for this example, the skewness factor was taken to be 1.0 for a perpendicular abutment.

Finally, for the purposes of comparison, it is reasonable to calculate the abutment scour for this case by using Froehlich's live-bed scour equation or the HIRE (Highways in the River Environment) equation as recommended by HEC-18. Substituting into Froehlich's equation, given previously as equation 5, and using the values of L_a, y_{f1}, and V_{f1} already determined, we have:

$$\frac{d_s}{y_{f_1}} = (2.27)(0.55)(1.0)(\frac{233}{0.458})^{0.43}(\frac{0.366}{\sqrt{(9.81)(0.458)}})^{0.61} + 1 = 7.24 \tag{63}$$

so that d_s = (7.24)(0.458) = 3.32 m (10.9 ft), which is approximately twice as large as the value estimated by the procedure proposed herein.

For the HIRE equation, which is recommended by HEC-18 for long abutments, the value of y_{ab} is taken to be 0.646 m (2.12 ft) as determined previously from the water-surface elevation in the bridge section. However, the value of V_{ab} is estimated to be equal to 0.98 m/s (3.2 ft/s) as determined from figure 30 or equation 59 developed in this research. *The use of the WSPRO results from the bridge section to estimate V_{ab} is not recommended because conveyance ratios cannot predict the local acceleration occurring near the abutment face.* Now, substituting into the HIRE equation given by equation 15 in chapter 2, we have:

$$\frac{d_s}{y_{ab}} = (4.0)(\frac{0.98}{\sqrt{(9.81)(0.646)}})^{1/3} = 2.9 \tag{64}$$

Then, if we add an FS of 1.0 for consistency with the other methods used, $d_s = (3.9)(0.646) = 2.52$ m (8.3 ft). This estimated scour depth lies between the one predicted from the results of this research (1.59 m) and the one from Froehlich's live-bed scour equation (3.32 m). The HIRE equation necessarily assumes live-bed scour, because there is no dependence on sediment size. In this example, live-bed scour is not predicted to occur.

COMPARISON WITH FIELD DATA

Field data for the flood of April 1997 on the Pomme de Terre River in Minnesota were measured by Mueller and Hitchcock[93] of the USGS and were provided to the author by Sterling Jones[94] of FHWA. A bridge scour investigation report for the bridge over Highway 12, including a WSPRO data file, was also made available.[95] Xibing Dou[96] provided his spreadsheets of the cross-sectional data for Highways 12 and 22 to the author.

New WSPRO runs were made by the author for each bridge, and scour calculations were carried out according to the methodology presented in this report. First, the downstream water-surface elevation at the exit section was adjusted for the measured discharge until the measured water-surface elevation at the bridge during the flood was reproduced. The bridge cross section after scour was used in these computer runs. Then, WSPRO runs were made using the cross sections before scour to obtain the necessary information for the scour predictions. The cross-section data for Highway 22 were more limited than that for Highway 12, so a template cross section was used to provide all of the necessary cross-section data in agreement with a measured river bed profile before scour.

Highway 22 Example

The Swift County Highway 22 bridge crosses the Pomme de Terre River near Artichoke Lake in Minnesota, about 10 km upstream from the Highway 12 bridge. The abutments are set at the edge of the main channel, and there are two bridge piers with a spacing of approximately 12.2 m (40.0 ft) as shown in figure 50. (The horizontal stations were established as increasing from right to left; however, the distances are given as negative numbers in the figure so that they are increasing from left to right while looking downstream.) The abutments are protected by rock riprap on a 2:1 sideslope. On April 5, 1997, a discharge of 129.5 m³/s (4570 ft³/s) was measured, and scour began to develop near the right abutment. On April 9, the discharge was measured as 145.9 m³/s (5150 ft³/s) and the bridge was temporarily closed until riprap could be placed to

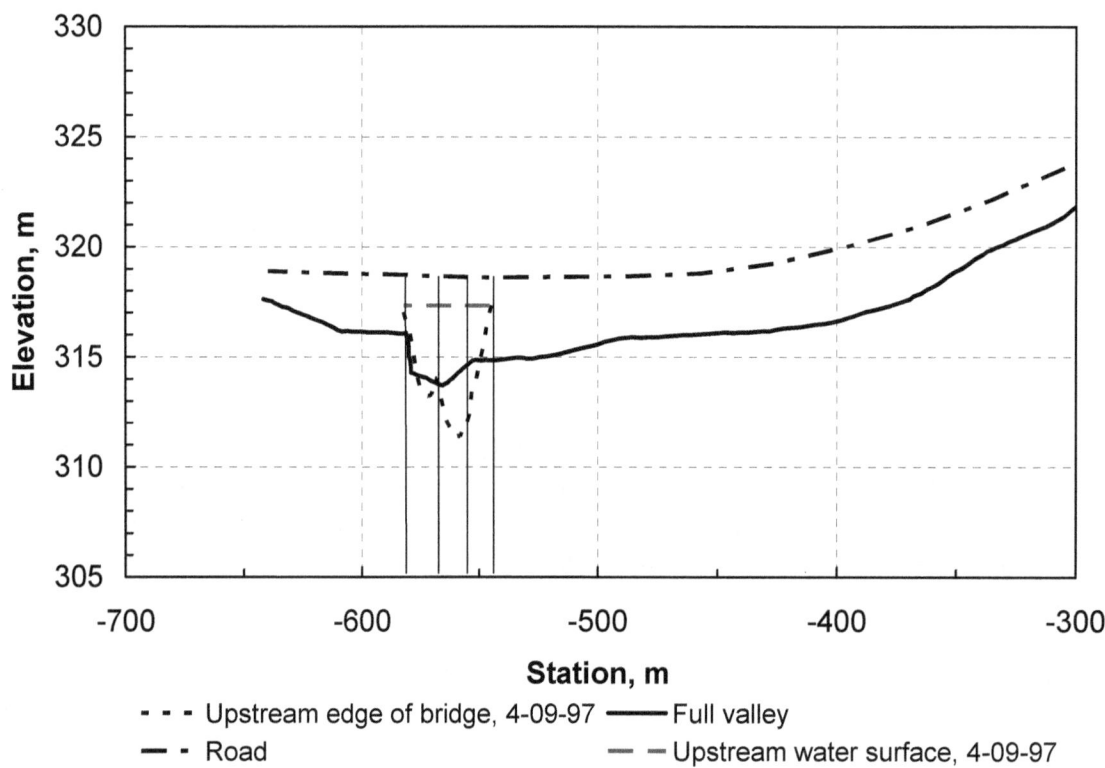

Figure 50. Cross sections for Highway 22 over the Pomme de Terre River
for the flood of April 9, 1997.

protect the bridge. The minimum bed elevation reached on that date was 311.3 m (1021.5 ft) as shown in figure 50. The bed material from elevation 312.4 m (1025.0 ft) downward is shown on the Minnesota Department of Highways bridge survey as being fine gravel.

From the after-scour WSPRO run for the measurements taken on April 9, it was determined that a water-surface elevation of 317.21 m (1040.96 ft) at the exit cross section gave an elevation of 317.25 m (1040.90 ft) at the downstream face of the bridge for Q = 145.9 m^3/s (5150 ft^3/s). Adding a drop of 0.09 m (0.30 ft) in the water surface through the bridge measured on April 4 at nearly the same discharge resulted in a water-surface elevation on the upstream face of the bridge of 317.34 m (1041.20 ft) in agreement with the measured value on April 9. The before-scour WSPRO run was then made with the same tailwater elevation at the exit section as for the after-scour run. The input data file is given as table 8, and the results are given in tables 9 through 11, from which the necessary information is obtained for the scour computations.

The value of the discharge distribution factor from table 9 is M = 1 – M(K) = 0.336.[1] The computations are made for the right abutment where the deepest scour occurred because of the wider floodplain on that side. The detailed procedure for making the computations has already been given in the previous example, so only the main results are highlighted in this field application. In table 9, the length of the abutment on the right side is determined to be 174.6 m (573 ft), and the water-surface elevation at the approach section can be obtained. From tables 10 and 11 for the velocity distributions at the approach section for the unconstricted and constricted water-surface elevations, respectively, it is determined that y_{f0} = 1.55 m (5.08 ft), V_{f1} = 0.24 m/s (0.79 ft/s), and y_{f1} = 1.67 m (5.48 ft). The median grain diameter d_{50} of the sediment is taken to be 2.0 mm, from which the critical velocity in the main channel of the bridge section is 0.87 m/s (2.85 ft/s) using Keulegan's equation, but substituting the hydraulic radius for the depth. From table 10, the velocities are all less than this critical velocity so that no sediment of the size being scoured could be carried into the scour hole (clear-water scour). The scour depth is then determined as clear-water scour around a bankline abutment from equation 56 using the shape factor for a spill-through abutment of 0.67, which is calculated from equation 57. The result is:

$$\frac{d_s}{y_{f0}} = 8.14 \ (0.67) \ [\frac{(0.24)(1.67)}{(0.336)(0.87)(1.55)} - 0.4] = 2.64 \tag{65}$$

The scour depth measured from the unconstricted water surface at the bridge is then $d_s + y_{f0}$ = (2.64)(1.55) + 1.55 = 5.6 m (18.4 ft) so that the bottom elevation in the scour hole is the water-surface elevation of 317.3 m (1041.0 ft) at the full valley section minus 5.6 m, or an elevation of 311.7 m (1022.7 ft). For comparison, the lowest measured bed elevation is 311.3 m (1021.4 ft), which is well within the standard error of the scour-prediction formula. Also, no FS has been included in the computation. If it is included, the predicted bottom elevation is 310.1 m (1017.4 ft), which provides a reasonable margin of safety.

---

[1] Alternatively, the value of M could be determined from the centerline of the main channel to the right edge of water in the approach section using the velocity distribution in Table 11, but it makes only a small difference in the result in this case.

Highway 12 Example

The second field example comes from the same flood on the Pomme de Terre River, but at the bridge on Highway 12 near Holloway, MN. The bridge was closed because of scour at the right abutment that extended below the abutment footing. The bridge is a single span with a length of 26.8 m (88.0 ft), and it has vertical abutments with 45-degree wingwalls. The channel and bridge cross sections are shown in figure 51. The measured discharge on April 9, 1997, was 162.9 m^3/s (5750 ft^3/s), with a measured water-surface elevation on the upstream side of the bridge of 302.6 m (993.0 ft). The minimum bed elevation was measured to be 293.5 m (963.0 ft) near the right abutment. The surficial sediment in the streambed is described as "organic, silty sand" with a d_{50} of 0.15 mm; however, the boring log shows that the material below about an elevation of 305 m (1000 ft) is "sand and gravel, gray, saturated."

From the after-scour WSPRO run, the exit tailwater elevation was determined to be 302.6 m (992.8 ft) for a discharge of 162.9 m^3/s (5750 ft^3/s). The input data file for the before-scour WSPRO run is given in table 12, and the WSPRO results are shown in tables 13 through 15. The value of the discharge distribution factor from table 13 is $M = 1 - M(K) = 0.335$. The computations are made for the right abutment (west) where the deepest scour occurred. From table 13, the length of the abutment on the right side is determined to be 121.3 m (398.0 ft), and the water-surface elevations at the approach section can be obtained. From tables 14 and 15 for the velocity distributions at the approach section for the unconstricted and constricted water-surface elevations, respectively, it is determined that $y_{f0} = 1.12$ m (3.67 ft), $V_{f1} = 0.19$ m/s (0.62 ft/s), and $y_{f1} = 1.22$ m (4.00 ft). If the median grain diameter d_{50} of the sediment is taken as 2 mm in the scour-hole region as done for Highway 22, the critical velocity in the main channel of the approach section is 0.88 m/s (2.89 ft/s) using Keulegan's equation. However, the WSPRO results in table 15 show maximum approach velocities of approximately 0.84 m/s (2.76 ft/s). For practical purposes, this means that the scour is approaching threshold live-bed scour. Thus, for live-bed scour around a bankline abutment, equation 58 is used with $V_{m1} = V_{m1c}$. Because M is relatively small, equation 58 indicates that the maximum value of $d_s/y_{f0} = 10$ has been reached. The scour depth measured from the water surface in the bridge before scour is then $d_s + y_{f0} = (10)(1.12) + 1.12 = 12.3$ m (40.4 ft), so that the bottom elevation in the scour hole is the unconstricted water-surface elevation of 302.6 m (992.9 ft) at the full valley section minus 12.3 m, or an elevation of 290.3 m (953 ft). For comparison, the lowest measured bed elevation is 293.5 m (963 ft), which is approximately 3 m higher than predicted. Given the presence of both sand and gravel, it is likely that armoring occurred, which limited the scour-hole depth in comparison to the predicted value; however, there is simply not enough information available on the sediment size distribution for this example, so these results are somewhat inconclusive.

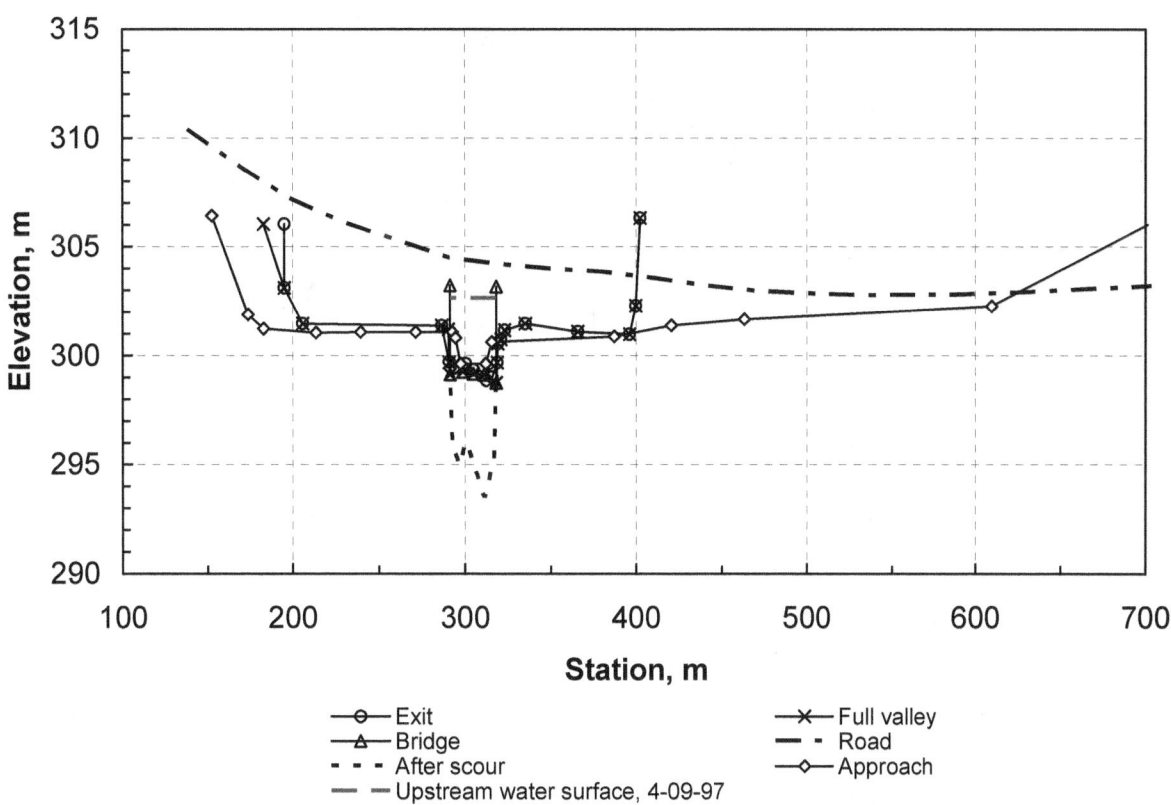

Figure 51. Cross sections for Highway 12 over the Pomme de Terre River for the flood of April 9, 1997.

LIMITATIONS

The experimental results developed herein are for the case of clear-water abutment scour in a compound channel with a wide floodplain. The ranges of variables covered in the experiments reported herein are given in table 3.

Table 3. Range of dimensionless variables
in experimental relationships.

Variable	Range
M	0.25-0.95
V_{fl}/V_{flc}	0.30-0.80
V_{f0}/V_{f0c}	0.40-1.00
y_{f0}/y_{fl}	0.40-1.00
L_a/B_f	0.17-1.00
$q_{fl}/(Mq_{f0c})$	0.40-2.30

CHAPTER 5. CONCLUSIONS AND RECOMMENDATIONS

The results of this research on bridge abutment scour in compound channels show that the discharge distribution factor is the appropriate variable to use rather than abutment length to measure the effects of flow contraction and flow redistribution in the contracted section on local scour depth. Sediment size is incorporated into the proposed scour-prediction technique in terms of the critical velocity required to initiate motion of the sediment. It is shown that the unconstricted flow depth in the floodplain should be used as the reference depth in the scour-prediction formula when bridge backwater is significant. The abutment shape is demonstrated to be important for shorter abutments, while no experimental differences in the abutment shape effects could be detected as the abutment increased in length and caused more contraction with encroachment on the main channel. Attempts to relate local abutment scour to local hydraulic variables near the abutment face were successful only for the shorter abutments; however, a relationship was developed between the approach hydraulic variables and the local contraction hydraulic variables. It is recommended that the proposed scour formula (equation 56) (which depends on the approach hydraulic variables predicted by WSPRO) be used. The suggested scour formula is for clear-water scour for both setback and bankline abutments. A second formula (equation 58) was proposed for bankline abutments that experience threshold live-bed scour conditions in the main channel; however, experiments in a compound section specifically designed for live-bed scour are needed. Testing of five current abutment scour formulas has shown that they all significantly overestimate or underestimate the scour data measured in this research. Field evaluation of the proposed method provided good agreement on one bridge, but an overestimate of scour on a second bridge. Armoring may have occurred in the latter case; however, inadequate data on the field sediment size distribution precluded a definite conclusion. It must be emphasized that the experimental results for bankline abutments that were used to develop the proposed scour formulas herein do not distinguish between contraction and abutment scour. Thus, the method of superposition of contraction and abutment scour for bankline abutments as though they were independent may be overly conservative.

It is concluded that the experimental results and methodology developed from this research can be used to estimate the depth of bridge abutment scour. It has been determined that for a compound channel over a much wider range of variables than was previously available that the effects of discharge distribution, sediment size, and time development on scour depth can be predicted from the relationships proposed herein. The comparisons of the WSPRO results, the experimental results, and the results from a two-dimensional numerical turbulence model have shown that the results from WSPRO are adequate for estimating the independent parameters needed for abutment scour prediction as long as bridge approach hydraulic conditions are used as predictor variables. It is recommended that the proposed procedure for abutment scour estimation be used alongside current FHWA procedures subject to the limitations on the ranges of dimensionless variables given in table 3, and that it be tested on additional field data as it becomes available. It is emphasized that sediment properties and their changes with depth must be known to adequately evaluate clear-water scour depths.

It is recommended that further research be conducted on abutment scour in order to evaluate and protect scour-critical bridges that are subject to possible failure. Suggested areas for research are:

1. Laboratory study of scour countermeasures is needed in order to design the most efficient abutment scour-protection schemes. The study should consider: (a) the extent, size, and placement of riprap at abutments, and (b) the effectiveness of spur dikes.

2. A three-dimensional numerical model with advanced turbulence closure schemes needs to be applied to the laboratory model used in this research for several selected cases of abutment scour. This effort should include three-dimensional velocity and turbulence measurements in the scour-hole area at different stages of scour-hole development with respect to time by temporarily fixing the bed in the scour-hole area.

3. A well-planned, detailed field study of a bridge subject to abutment scour in cooperation with the USGS is needed. The bridge should be instrumented and scour determined over a 3-year period with detailed field measurements of velocity and bed elevation changes. A laboratory model of the instrumented field bridge should be constructed and tested to compare scour predictions based on the laboratory data with those actually measured in the field to finally resolve laboratory scale-up issues.

4. Further experimental investigation of the live-bed scour case with the abutment at the edge of the main channel is warranted. This will require special design of the compound-channel geometry so that a careful selection of a combination of flow rates, sediment size, and discharge distribution will result in sediment transport in the main channel without scour exceeding the physical limits imposed by the finite depth of the sediment.

APPENDIX A.
WSPRO INPUT AND OUTPUT FOR BURDELL CREEK, HIGHWAY 22 BRIDGE OVER POMME DE TERRE RIVER, AND HIGHWAY 12 BRIDGE OVER POMME DE TERRE RIVER

Table 4. WSPRO input data file for Burdell Creek, Q_{100}.

```
*F
SI 0
T1 ABUTMENT SCOUR EXAMPLE
*
Q  14000
SK 0.001
*
XS EXIT 750 0.0 0.5 0.0 0.001
GR   0, 19    100, 15    200, 11    500, 10.75    900, 10    1100, 9.0
GR   1215, 5.5    1250, 4.9    1300, 3.05    1350, 4.85    1385, 5.1
GR   1500, 9.0    1700, 10    2100, 10.75    2400, 11    2500, 15
GR   2600, 19
N  0.042 0.032 0.042
SA 1100 1500
*
XS FULV 1500 0.0 0.5 0.0
*
BR BRDG 1500 * 0.0 0.5 0.0 *
BL 0 750 1100 1500
BC 18
CD 3 50 2 22 0.0 0.0 0.0
AB 2 2
PD 0  5.65  30  6
N  0.042 0.032  0.042
SA 1100  1500
*
XS APPR 2300 0.0 0.5 0.0
*
HP 2    APPR  13.25   0  13.25  14000
HP 2    APPR  13.64   0  13.64  14000
*
DA BRDG  1  1  *  *  14000  1
*
EX 0
ER
```

Table 5. WSPRO water-surface profile output for Burdell Creek, Q_{100}.

```
************************* W S P R O ***************************
         Federal Highway Administration  -  U.S. Geological Survey
                  Model for Water-Surface Profile Computations.
                  Input Units: English  /  Output Units: English
    *-----------------------------------------------------------------*
                           ABUTMENT SCOUR EXAMPLE

                    WSEL    VHD       Q          AREA        SRDL        LEW
                    EGEL    HF        V           K          FLEN        REW
                    CRWS    HO       FR #         SF         ALPHA       ERR
                  --------- ------ ---------- ---------- --------- ---------
Section: EXIT      11.696   .268  14000.000   4721.088  ********   182.604
Header Type: XS    11.964  ******     2.965  442473.00  ********  2417.396
SRD:    750.000     9.447  ******      .504    ******      1.960   ******


Section: FULV      12.449   .267  14000.000   4728.060   750.000   182.526
Header Type: FV    12.716   .750      2.961  443183.70   750.000  2417.474
SRD:   1500.000    10.197   .000       .502      .0010     1.959      .003

        <<< The Preceding Data Reflect The "Unconstricted" Profile >>>

Section: APPR      13.250   .267  14000.000   4730.357   800.000   182.500
Header Type: AS    13.517   .798      2.960  443418.00   800.000  2417.500
SRD:   2300.000    10.997   .000       .502      .0010     1.959      .003

        <<< The Preceding Data Reflect The "Unconstricted" Profile >>>
        <<< The Following Data Reflect The "Constricted" Profile >>>
          <<< Beginning Bridge/Culvert Hydraulic Computations >>>

                    WSEL    VHD       Q          AREA        SRDL        LEW
                    EGEL    HF        V           K          FLEN        REW
                    CRWS    HO       FR #         SF         ALPHA       ERR
                  --------- ------ ---------- ---------- --------- ---------
Section: BRDG      12.470   .470  14000.000   3038.370   750.000   936.685
Header Type: BR    12.940   .921      4.608  386306.30   750.000  1663.315
SRD:   1500.000    10.668   .055       .474     ******     1.424     -.004

Specific Bridge Information    C     P/A    PFELEV    BLEN      XLAB     XRAB
Bridge Type 3   Flow Type 1 ------  -----  --------  --------  --------  -------
Pier/Pile Code  0            .8380   .067    18.000   750.000  940.750  659.250
--------------------------- ------  -----  --------  --------  --------  -------

                    WSEL    VHD       Q          AREA        SRDL        LEW
                    EGEL    HF        V           K          FLEN        REW
                    CRWS    HO       FR #         SF         ALPHA       ERR
                  --------- ------ ---------- ---------- --------- ---------
Section: APPR      13.635   .186  14000.000   5595.480   750.000   172.865
Header Type: AS    13.822   .780      2.502  536819.40   776.571  2427.135
SRD:   2300.000    10.997   .100       .387      .0010     1.914     -.017

          Approach Section APPR Flow Contraction Information
             M( G )   M( K )    KQ       XLKQ      XRKQ      OTEL
           -------- -------- --------- -------- -------- --------
              .675     .189  438924.8  938.577  1661.368  13.635
```

Table 6. Velocity distribution from HP record for unconstricted flow at approach section, Burdell Creek, Q_{100}.

```
************************  W S P R O  **************************
        Federal Highway Administration  -  U.S. Geological Survey
               Model for Water-Surface Profile Computations.
               Input Units: English  /  Output Units: English
   *----------------------------------------------------------------*
                          ABUTMENT SCOUR EXAMPLE


       ***    Beginning Velocity Distribution For Header Record APPR    ***
              SRD Location:   2300.000        Header Record Number  4

           Water-Surface Elevation:      13.250              Element # 1
            Flow:  14000.000   Velocity: 2.96 Hydraulic Depth:  2.116
            Cross-Section Area:   4730.38        Conveyance:   443419.90
              Bank Stations ->  Left:   182.500   Right:  2417.500

X STA.       182.5        803.7      1045.3      1141.7      1181.9      1209.7
   A( I )               628.6       454.8       279.4       184.1       156.0
   V( I )                1.11        1.54        2.51        3.80        4.49
   D( I )                1.01        1.88        2.90        4.58        5.62

X STA.      1209.7       1232.9      1253.5      1271.9      1288.1      1302.5
   A( I )               146.1       138.2       133.4       128.5       121.2
   V( I )                4.79        5.07        5.25        5.45        5.77
   D( I )                6.30        6.69        7.27        7.91        8.46

X STA.      1302.5       1317.4      1333.9      1352.7      1373.2      1394.8
   A( I )               124.2       127.1       133.8       138.2       141.8
   V( I )                5.64        5.51        5.23        5.07        4.94
   D( I )                8.29        7.73        7.10        6.76        6.55

X STA.      1394.8       1421.9      1458.5      1558.5      1799.0      2417.5
   A( I )               157.2       172.9       290.7       449.8       624.5
   V( I )                4.45        4.05        2.41        1.56        1.12
   D( I )                5.81        4.73        2.91        1.87        1.01
```

Table 7. Velocity distribution from HP record for constricted flow at approach section, Burdell Creek, Q$_{100}$.

```
*********************** W S P R O ***************************
      Federal Highway Administration  -  U.S. Geological Survey
            Model for Water-Surface Profile Computations.
            Input Units: English  /  Output Units: English
    *----------------------------------------------------------------*
                        ABUTMENT SCOUR EXAMPLE


    ***    Beginning Velocity Distribution For Header Record APPR    ***
           SRD Location:   2300.000      Header Record Number  4

       Water-Surface Elevation:    13.640              Element # 1
        Flow:   14000.000   Velocity:  2.50  Hydraulic Depth:  2.487
        Cross-Section Area:    5605.83        Conveyance:   537996.80
          Bank Stations ->  Left:   172.750   Right:  2427.250
```

X STA.	172.7	684.7	945.3	1101.6	1161.2	1195.4
A(I)		658.8	506.4	423.1	241.3	187.3
V(I)		1.06	1.38	1.65	2.90	3.74
D(I)		1.29	1.94	2.71	4.05	5.47

X STA.	1195.4	1223.0	1246.6	1268.0	1286.0	1302.5
A(I)		176.6	163.5	159.7	147.8	144.8
V(I)		3.96	4.28	4.38	4.74	4.83
D(I)		6.40	6.93	7.47	8.19	8.81

X STA.	1302.5	1319.1	1338.0	1359.2	1382.3	1407.8
A(I)		143.9	151.0	156.4	163.7	169.5
V(I)		4.87	4.64	4.48	4.28	4.13
D(I)		8.65	8.01	7.35	7.09	6.64

X STA.	1407.8	1441.5	1497.0	1658.1	1917.6	2427.3
A(I)		190.1	229.2	435.6	502.4	655.0
V(I)		3.68	3.05	1.61	1.39	1.07
D(I)		5.64	4.13	2.70	1.94	1.29

Table 8. WSPRO input data file for Highway 22 bridge.

```
*F
SI 0
T1                    BRIDGE 76518
T2            HIGHWAY 22 OVER POMME DE TERRE RIVER
T3            EXISTING BRIDGE, 1997 FLOOD ANALYSIS, BEFORE SCOUR
J1 0.25
Q  5150
WS 1040.96
*  TEMPLATE SECTION AT BRIDGE, FULL VALLEY
XT FV      1020
GR -2106.1,1042.1 -2094.9,1041.8 -2090.5,1041.5 -2080.7,1041.0 -2040.1,1039.3
GR -2003.4,1037.5 -1916.3,1037.0 -1911.3,1037.0 -1907.3,1037.1 -1905.9,1036.3
GR -1904.1,1035.8 -1903.5,1035.1 -1903.4,1034.4 -1903.3,1033.7 -1901.3,1032.9
GR -1900.5,1032.0 -1900.3,1031.2 -1897.1,1031.1 -1891.6,1030.8 -1885.6,1030.6
GR -1878.8,1030.4 -1875.8,1030.2 -1871.2,1029.9 -1866.8,1029.7 -1861.7,1029.4
GR -1855.9,1029.3 -1849.3,1029.7 -1845.9,1030.1 -1841.9,1030.4 -1834.4,1031.2
GR -1829.6,1031.6 -1825.4,1032.0 -1819,1032.5   -1813.8,1033   -1778.3,1033.1
GR -1757.4,1033.4 -1730.9,1033.3 -1700.4,1033.8 -1651.6,1035.1 -1603.5,1036.4
GR -1546.7,1036.6 -1503.1,1036.9 -1405.6,1037.4 -1351.8,1038.2 -1329.4,1038.5
GR -1318.1,1038.7 -1307.3,1039   -1298.7,1039.3 -1289.8,1039.6 -1278.8,1040
GR -1271.4,1040.3 -1259.1,1040.6 -1248.1,1040.9 -1238.3,1041.2 -1229.4,1041.5
GR -1219.6,1041.8 -1211.8,1042.1 -1206.6,1042.6 -1197.7,1042.9 -1188.1,1043.6
*
XS EXIT    900
GT +0.60
N   0.08     0.035     0.08
SA      -1910     -1790
*
XS FULV    1020    *       *       *   -0.005
*
BR BRGE    1020   1041.21
GR -1910,1041.2    -1908,1039.2    -1903,1037.7   -1897,1034    -1891.6,1030.8
GR -1885.6,1030.6 -1878.8,1030.4 -1875.8,1030.2 -1871.2,1029.9 -1866.8,1029.7
GR -1861.7,1029.4 -1855.9,1029.3 -1849.3,1029.7 -1845.9,1030.1 -1841.9,1030.4
GR -1834.4,1031.2 -1829.6,1031.6 -1825.4,1032   -1819,1032.5   -1813.8,1033
GR -1803,1033.5    -1798,1035.7    -1793,1039.2   -1790,1041.2   -1910,1041.2
N  0.035
*
PW 0  1029.8,1.5   1031.7,1.5    1031.7,3
*  TYPE        BRWDTH     EMBSS      EMBELEV
CD 3            40          2          1045.4
KQ *      *      *       -1910    -1790
*
HP 2 BRGE      1040.82     *       *          5150
*
XS APPR       1180
N  0.05      0.035     0.05
SA      -1910     -1790
*
HP 2  APPR  1041.08     *       *       5150
HP 2  APPR  1041.56     *       *       5150
*
EX
ER
```

113

Table 9. WSPRO water-surface profile output for Highway 22 bridge.

```
************************  W S P R O  ***************************
        Federal Highway Administration  -  U.S. Geological Survey
              Model for Water-Surface Profile Computations.
              Input Units: English  /  Output Units: English
    *-----------------------------------------------------------*
                            BRIDGE 76518

                 HIGHWAY 22 OVER POMME DE TERRE RIVER
             EXISTING BRIDGE, 1997 FLOOD ANALYSIS, BEFORE SCOUR
```

	WSEL	VHD	Q	AREA	SRDL	LEW
	EGEL	HF	V	K	FLEN	REW
	CRWS	HO	FR #	SF	ALPHA	ERR
	---------	------	---------	----------	---------	---------
Section: EXIT	1040.960	.097	5150.000	3511.259	*********	-2065.417
Header Type: XS	1041.057	******	1.467	298703.30	*********	-1268.937
SRD: 900.000	1035.804	******	.210	******	2.894	******
Section: FULV	1041.014	.073	5150.000	4045.370	120.000	-2080.970
Header Type: FV	1041.087	.030	1.273	352758.00	120.000	-1244.383
SRD: 1020.000	1035.204	.000	.174	.0003	2.908	.000

```
        <<< The Preceding Data Reflect The "Unconstricted" Profile >>>

    ===135 CONVEYANCE RATIO OUTSIDE OF RECOMMENDED LIMITS AT "APPR".
          KRATIO:  1.54
```

	WSEL	VHD	Q	AREA	SRDL	LEW
Section: APPR	1041.079	.030	5150.000	4787.963	160.000	-2097.845
Header Type: AS	1041.109	.022	1.076	544559.30	160.000	-1217.549
SRD: 1180.000	1034.376	.000	.106	.0001	1.684	.000

```
        <<< The Preceding Data Reflect The "Unconstricted" Profile >>>

        <<< The Following Data Reflect The "Constricted" Profile >>>
           <<< Beginning Bridge/Culvert Hydraulic Computations >>>

    ===220 FLOW CLASS 1 ( 4 ) SOLUTION INDICATES POSSIBLE PRESSURE FLOW.
          WS3, WSIU, WS1, PFELV:  1040.82   1041.52   1041.55   1041.21

    ===245 ATTEMPTING FLOW CLASS 2 ( 5 ) SOLUTION.

    ===250 INSUFFICIENT HEAD FOR PRESSURE FLOW.
          YU/Z, WSIU, WS:  1.06   1041.70   1041.76

    ===270 REJECTED FLOW CLASS 2 ( 5 ) SOLUTION.
```

	WSEL	VHD	Q	AREA	SRDL	LEW
	EGEL	HF	V	K	FLEN	REW
	CRWS	HO	FR #	SF	ALPHA	ERR
	---------	------	---------	----------	---------	---------
Section: BRGE	1040.825	.673	5150.000	1014.368	120.000	-1909.625
Header Type: BR	1041.498	.063	5.077	174703.30	120.000	-1790.563
SRD: 1020.000	1035.708	.378	.397	******	1.678	.000

Table 9. WSPRO water-surface profile output for Highway 22 bridge (continued).

```
Specific Bridge Information   C     P/A    PFELEV    BLEN      XLAB      XRAB
Bridge Type 3   Flow Type 1 ------ ----- -------- -------- -------- -------
Pier/Pile Code  0             .7719 .030  1041.210 ******** ******** *******
--------------------------- ------ ----- -------- -------- -------- -------


                    WSEL    VHD      Q         AREA     SRDL      LEW
                    EGEL    HF       V          K       FLEN      REW
                    CRWS    HO      FR #        SF      ALPHA     ERR
                    --------- ------ ---------- ---------- --------- ---------
Section: APPR       1041.555 .025   5150.000   5212.262  120.000 -2106.100
Header Type: AS     1041.580 .038       .988   611618.10 149.347 -1209.151
SRD:    1180.000    1034.376 .044       .093       .0001   1.647      .000

           Approach Section APPR Flow Contraction Information
           M( G )   M( K )     KQ      XLKQ     XRKQ      OTEL
           -------- -------- --------- -------- -------- --------
             .864      .664  205746.2 ******** ******** 1041.555
           -------- -------- --------- -------- -------- --------

           <<< End of Bridge Hydraulics Computations >>>
```

Table 10. Velocity distribution from HP record for unconstricted flow at approach section, Highway 22 bridge.

```
*********************** W S P R O ***************************
      Federal Highway Administration  -  U.S. Geological Survey
            Model for Water-Surface Profile Computations.
            Input Units: English  /  Output Units: English
    *------------------------------------------------------------*
                          BRIDGE 76518

                 HIGHWAY 22 OVER POMME DE TERRE RIVER

              EXISTING BRIDGE, 1997 FLOOD ANALYSIS, BEFORE SCOUR


      ***     Beginning Velocity Distribution For Header Record APPR     ***
              SRD Location:   1180.000       Header Record Number  4

         Water-Surface Elevation:   1041.080              Element # 1
         Flow:    5150.000   Velocity:  1.08  Hydraulic Depth:  5.440
         Cross-Section Area:  4788.92          Conveyance:   544702.60
         Bank Stations -> Left: -2097.886   Right: -1217.521

    X STA.    -2097.9     -1951.7    -1898.1    -1885.5    -1874.3    -1864.0
      A( I )                439.4      286.6      138.3      129.1      124.7
      V( I )                 .59        .90       1.86       1.99       2.07
      D( I )                3.01       5.34      11.04      11.48      12.07

    X STA.    -1864.0     -1854.1    -1843.7    -1832.0    -1817.9    -1800.8
      A( I )                123.3      125.6      129.5      139.2      151.8
      V( I )                2.09       2.05       1.99       1.85       1.70
      D( I )               12.50      12.08      11.05       9.87       8.91

    X STA.    -1800.8     -1780.6    -1753.6    -1725.3    -1694.6    -1658.4
      A( I )                178.8      232.2      241.8      252.8      269.2
      V( I )                1.44       1.11       1.06       1.02        .96
      D( I )                8.81       8.62       8.54       8.23       7.44

    X STA.    -1658.4     -1611.9    -1551.5    -1484.9    -1404.4    -1217.5
      A( I )                294.6      327.1      339.1      376.9      488.9
      V( I )                 .87        .79        .76        .68        .53
      D( I )                6.34       5.42       5.09       4.68       2.62
```

Table 11. Velocity distribution from HP record for constricted flow at approach section, Highway 22 bridge.

```
*********************** W S P R O **************************
    Federal Highway Administration - U.S. Geological Survey
          Model for Water-Surface Profile Computations.
         Input Units: English / Output Units: English
 *---------------------------------------------------------*
                        BRIDGE 76518

          HIGHWAY 22 OVER POMME DE TERRE RIVER

       EXISTING BRIDGE, 1997 FLOOD ANALYSIS, BEFORE SCOUR


   ***    Beginning Velocity Distribution For Header Record APPR    ***
         SRD Location:   1180.000      Header Record Number  4

      Water-Surface Elevation:   1041.560           Element # 1
      Flow:   5150.000   Velocity:  .99 Hydraulic Depth:  5.816
      Cross-Section Area:  5217.09       Conveyance:  612424.90
        Bank Stations -> Left: -2106.100  Right: -1209.095

X STA.   -2106.1    -1958.7    -1899.9    -1886.7    -1874.7    -1863.8
  A( I )             476.5      327.6      152.2      142.2      137.3
  V( I )              .54        .79       1.69       1.81       1.88
  D( I )             3.23       5.57      11.46      11.93      12.54

X STA.   -1863.8    -1853.3    -1842.5    -1830.0    -1814.5    -1796.9
  A( I )             135.7      134.8      142.0      157.3      164.5
  V( I )             1.90       1.91       1.81       1.64       1.57
  D( I )            12.98      12.48      11.38      10.12       9.34

X STA.   -1796.9    -1774.0    -1745.5    -1716.7    -1684.8    -1645.8
  A( I )             212.0      257.4      258.9      272.6      296.9
  V( I )             1.21       1.00        .99        .94        .87
  D( I )             9.28       9.04       8.99       8.53       7.62

X STA.   -1645.8    -1595.5    -1536.5    -1470.1    -1392.8    -1209.1
  A( I )             324.2      343.6      363.4      393.1      524.8
  V( I )              .79        .75        .71        .66        .49
  D( I )             6.44       5.83       5.48       5.08       2.86
```

Table 12. WSPRO input data file for Highway 12 bridge.

```
*F
SI 0
T1                        BRIDGE 5359
T2                TH 12 OVER POMME DE TERRE RIVER
T3          ****   EXISTING BRIDGE BEFORE SCOUR, 1997 FLOOD ANALYSIS      ****
J1        0.25
Q         5750
WS        992.8
*            FIELD SURVEYED X-SEC. 75' DOWNSTREAM FROM BRIDGE 5359
XS   DS100  25
GR        640,1004.09  640,994.49  675,989.09
*         750,987.19  800,1005.0   800,989.29   850,989.19   900,989.89
GR        940,988.79   953,988.19
GR        954,983.36   965,982.56   985,983.06  1000,982.16  1025,980.56
GR       1036,981.06 1046,983.26  1047,986.09  1053,986.69  1060,988.09
GR       1100,989.09 1200,987.89  1300,987.49  1312,991.79  1320,1005.0
N         0.08    0.03    0.08
SA              953     1047
*            FULL VALLEY X-SEC. COMPOSED FROM PROPAGATING XS-DS100 UP TO
*            BRIDGE AND SUPERIMPOSING THE SURVEYED BRIDGE OPENING
XS   FULLV  100
GR        600,1004.09  640,994.49  675,989.09
GR        940,988.79   953,988.19
GR        954,983.26   956,981.48   980,981.85  1000,981.54  1020,981.39
GR   .   1044,980.19  1046,983.26  1047,986.09  1053,986.69  1060,988.09
GR       1100,989.09 1200,987.89  1300,987.49  1312,991.79  1320,1005.0
*            BRIDGE DEFINITION RECORDS FOLLOW
BR   BRDGE  100, 994.86
GR        956,994.86   956,981.48   980,981.85  1000,981.54
GR       1020,981.39  1044,980.19  1044,994.67   956,994.86
N             0.035
*        BRTYPE  BRWIDTH  EMBSS  EMBELV   WWANGL   WWWID
CD        4        51       3      998      45
KQ        *,*,*, 956, 1044
*
XR   ROAD   125, 50
GR        454,1018.39  554,1012.83  654,1007.87  754,1004.34  854,1001.74
GR        954,999.09  1045,998.18  1146,997.44  1246,996.98  1346,996.03
GR       1446,994.89  1546,994.09  1646,993.71  1746,993.45  1846,993.50
GR       1946,993.55  2046,993.83  4689,1003.83
*  IHP SEC  ELMIN     YINC      ELMAX       Q
HP 2 BRDGE  992.63      *         *         5750
AS   APPR   325
GR        500,1005.36  570,990.45   600,988.35
GR        700,987.75   785,987.85   890,987.8   960,987.8    966,986.95
GR        976,983.09  1024,983.09  1035,986.35 1270,987.15  1380,988.85
GR       1520,989.75  2000,991.75  2300,1004.0
N         0.080    0.05    0.030      0.05       0.080
SA              950     966      1035     1100
*  IHP SEC   ELMIN     YINC    ELMAX      Q
HP 2  APPR   993.05      *        *       5750
HP 2  APPR   993.40      *        *       5750
EX
ER
```

118

Table 13. WSPRO water-surface profile output for Highway 12 bridge.

```
************************  W S P R O  ***************************
     Federal Highway Administration  -  U.S. Geological Survey
          Model for Water-Surface Profile Computations.
          Input Units: English  /  Output Units: English
*----------------------------------------------------------*

                       BRIDGE 5359


              TH 12 OVER POMME DE TERRE RIVER
     ****    EXISTING BRIDGE BEFORE SCOUR, 1997 FLOOD ANALYSIS    ****
```

| | WSEL | VHD | Q | AREA | SRDL | LEW |
| | EGEL | HF | V | K | FLEN | REW |
	CRWS	HO	FR #	SF	ALPHA	ERR
Section: DS100	992.800	.245	5750.000	2281.291	*********	940.000
Header Type: XS	993.045	******	2.521	295141.50	*********	1312.612
SRD: 25.000	987.030	******	.283	******	2.483	******
Section: FULLV	992.923	.145	5750.000	3492.329	75.000	650.158
Header Type: FV	993.068	.023	1.646	373264.60	75.000	1312.686
SRD: 100.000	986.366	.000	.234	.0003	3.435	.000

```
<<< The Preceding Data Reflect The "Unconstricted" Profile >>>
```

	WSEL	VHD	Q	AREA	SRDL	LEW
Section: APPR	993.048	.048	5750.000	6355.054	225.000	557.801
Header Type: AS	993.097	.046	.905	437889.90	225.000	2031.796
SRD: 325.000	989.048	.000	.150	.0002	3.791	-.017

```
<<< The Preceding Data Reflect The "Unconstricted" Profile >>>
<<< The Following Data Reflect The "Constricted" Profile >>>
   <<< Beginning Bridge/Culvert Hydraulic Computations >>>
```

| | WSEL | VHD | Q | AREA | SRDL | LEW |
| | EGEL | HF | V | K | FLEN | REW |
	CRWS	HO	FR #	SF	ALPHA	ERR
Section: BRDGE	992.637	.647	5750.000	989.807	75.000	956.017
Header Type: BR	993.284	.052	5.809	180129.70	75.000	1044.086
SRD: 100.000	986.485	.185	.339	******	1.233	-.017

Specific Bridge Information	C	P/A	PFELEV	BLEN	XLAB	XRAB
Bridge Type 4 Flow Type 1	------	-----	--------	--------	--------	-------
Pier/Pile Code **	.9006	.000	994.860	********	********	*******

Table 13. WSPRO water-surface profile output for Highway 12 bridge (continued).

```
                *** Roadway Section Located at SRD   125.000 ***

                      Section:  ROAD       Header Type: XR
                      <<< Embankment Is Not Overtopped >>>

                    WSEL      VHD        Q         AREA       SRDL       LEW
                    EGEL      HF         V          K         FLEN       REW
                    CRWS      HO        FR #        SF        ALPHA      ERR

                   --------- ------ ---------- ---------- --------- ---------
Section: APPR       993.401   .040   5750.000   6876.729   174.000   556.146
Header Type: AS     993.441   .093      .836    487502.70  222.813  2040.433
SRD:     325.000    989.048   .066      .131       .0002     3.638      .013

               Approach Section APPR  Flow Contraction Information
               M( G )   M( K )    KQ        XLKQ      XRKQ      OTEL
              -------- -------- --------- -------- -------- --------
                .940     .665   162782.2  971.035 1059.104  993.377
              -------- -------- --------- -------- -------- --------

               <<< End of Bridge Hydraulics Computations >>>
```

120

Table 14. Velocity distribution from HP record for unconstricted flow at approach section, Highway 12 bridge.

```
************************ W S P R O ***************************
       Federal Highway Administration  -  U.S. Geological Survey
             Model for Water-Surface Profile Computations.
             Input Units: English  /  Output Units: English
*------------------------------------------------------------------*
                         BRIDGE 5359

                  TH 12 OVER POMME DE TERRE RIVER
         ****    EXISTING BRIDGE BEFORE SCOUR, 1997 FLOOD ANALYSIS   ****


      ***    Beginning Velocity Distribution For Header Record APPR    ***
             SRD Location:   325.000      Header Record Number  5

        Water-Surface Elevation:   993.050           Element # 1
         Flow:   5750.000  Velocity:  .90 Hydraulic Depth:  4.313
         Cross-Section Area:  6357.50       Conveyance:   438116.30
             Bank Stations -> Left:   557.794  Right:  2031.836

 X STA.     557.8      668.9      742.2      819.3      893.2      961.6
   A( I )             463.3      384.6      402.5      386.9      359.0
   V( I )              .62        .75        .71        .74        .80
   D( I )             4.17       5.25       5.22       5.23       5.25

 X STA.     961.6      977.9      987.5      997.2     1006.8     1016.5
   A( I )             124.5       96.1       96.8       95.7       95.7
   V( I )             2.31       2.99       2.97       3.00       3.00
   D( I )             7.64       9.96       9.96       9.96       9.96

 X STA.    1016.5     1026.4     1047.5     1079.6     1121.4     1182.3
   A( I )              98.3      152.2      211.7      270.9      383.8
   V( I )             2.93       1.89       1.36       1.06        .75
   D( I )             9.87       7.21       6.60       6.48       6.30

 X STA.    1182.3     1249.8     1329.2     1451.5     1635.2     2031.8
   A( I )             410.3      442.1      517.5      593.6      772.2
   V( I )              .70        .65        .56        .48        .37
   D( I )             6.08       5.57       4.23       3.23       1.95
```

Table 15. Velocity distribution from HP record for constricted flow at approach section, Highway 12 bridge.

```
*************************  W S P R O  **************************
       Federal Highway Administration  -  U.S. Geological Survey
            Model for Water-Surface Profile Computations.
            Input Units: English  /  Output Units: English
  *-------------------------------------------------------------*
                          BRIDGE 5359

                 TH 12 OVER POMME DE TERRE RIVER
        ****   EXISTING BRIDGE BEFORE SCOUR, 1997 FLOOD ANALYSIS   ****


     ***    Beginning Velocity Distribution For Header Record APPR    ***
            SRD Location:   325.000      Header Record Number  5

        Water-Surface Elevation:   993.400           Element # 1
        Flow:   5750.000  Velocity:  .84  Hydraulic Depth:  4.632
        Cross-Section Area:  6875.20      Conveyance:  487355.20
           Bank Stations -> Left:   556.150  Right:   2040.409

X STA.    556.2      666.3      740.6      816.6      892.3      960.4
  A( I )             488.6      415.4      423.2      422.7      381.0
  V( I )              .59        .69        .68        .68        .75
  D( I )             4.43       5.59       5.57       5.58       5.60

X STA.    960.4      978.1      988.2      998.4     1008.5     1018.5
  A( I )             139.4      104.3      105.1      103.8      103.8
  V( I )             2.06       2.76       2.74       2.77       2.77
  D( I )             7.87      10.31      10.31      10.31      10.31

X STA.   1018.5     1029.4     1056.2     1089.4     1139.5     1202.8
  A( I )             107.8      192.5      229.9      339.5      417.2
  V( I )             2.67       1.49       1.25        .85        .69
  D( I )             9.91       7.19       6.92       6.78       6.59

X STA.   1202.8     1272.2     1356.7     1489.1     1673.2     2040.4
  A( I )             441.1      470.3      568.1      626.2      795.2
  V( I )              .65        .61        .51        .46        .36
  D( I )             6.36       5.56       4.29       3.40       2.17
```

REFERENCES

1. Davis, S.R. (1984). "Case Histories of Scour Problems at Bridges," *Transportation Research Record*, Transportation Research Board (TRB), Washington, DC, 950(2), pp. 149-155.

2. Lagasse, P.F.; Schall, J.D.; and Frick, D.M. (1988). "Analytical Studies of the Schoharie Bridge Failure," *Proceedings of the National Conference on Hydraulic Engineering*, Colorado Springs, CO, ASCE, pp. 521-527.

3. Richardson, E.V.; Ruff, J.F.; and Brisbane, T.E. (1988). "Schoharie Creek Bridge Model Study," *Proceedings of the National Conference on Hydraulic Engineering*, Colorado Springs, CO, ASCE, pp. 528-533.

4. Richardson, E.V. (1996). "Historical Development of Bridge Scour Evaluations," *Proceedings of the North American Water and Environment Congress* (on CD-ROM), Anaheim, CA, ASCE.

5. Parola, A.C.; Hagerty, D.J.; Mueller, D.S.; Melville, B.W.; Parker, G.; and Usher, J.S. (1997). "The Need for Research on Scour at Bridge Crossings," *Managing Water: Coping With Scarcity and Abundance*, 27th Congress of the International Association for Hydraulic Research, San Francisco, CA, ASCE, pp. 124-129.

6. Stamey, Timothy C. (1996). Summary of Data-Collection Activities and Effects of Flooding From Tropical Storm Alberto in Parts of Georgia, Alabama, and Florida, July 1994, Open-File Report 96-228, USGS, Reston, VA, 23 pp.

7. Pagan-Ortiz, J.E. (1998). "Status of the Scour Evaluation of Bridges Over Waterways in the United States," *Proceedings of Water Resources Engineering, '98*, Memphis, TN, ASCE, pp. 2-4.

8. Richardson, E.V., and Davis, Stan (1995). *Evaluating Scour at Bridges*, HEC-18, FHWA, USDOT, Washington, DC.

9. Richardson, E.V., and Richardson, J.R. (1998). Discussion of "Pier and Abutment Scour: Integrated Approach," by B.W. Melville, *Journal of Hydraulic Engineering*, ASCE, 124(7), pp. 771-772.

10. Richardson, J.R., and Richardson, E.V. (1993). Discussion of "Local Scour at Bridge Abutments," by B.W. Melville, *Journal of Hydraulic Engineering*, ASCE, 119(9), pp. 1069-1071.

11. Sturm, T.W. (1999). "Abutment Scour in Compound Channels," *Stream Stability and Scour at Highway Bridges*, edited by E.V. Richardson and P.F. Lagasse, ASCE, pp. 443-456.

12. Froehlich, D.C. (1989). "Local Scour at Bridge Abutments," *Proceedings of the National Conference on Hydraulic Engineering*, New Orleans, LA, ASCE, pp. 13-18.

13. Laursen, E.M. (1963). "An Analysis of Relief Bridge Scour," *Journal of the Hydraulics Division*, ASCE, 92(HY3), pp. 93-118.

14. Melville, B.W. (1992). "Local Scour at Bridge Abutments," *Journal of Hydraulic Engineering*, ASCE, 118(4), pp. 615-631.

15. Melville, B.W. (1995). "Bridge Abutment Scour in Compound Channels," *Journal of Hydraulic Engineering*, ASCE, 121(12), pp. 863-868.

16. Sturm, T.W., and Sadiq, Aftab (1991). "Water Surface Profiles and the Compound Channel Froude Number for Rough Floodplains," *Proceedings of the International Symposium on Environmental Hydraulics*, Hong Kong, edited by J.H.W. Lee and Y.K. Cheung, A.A. Balkema Publishers, Rotterdam, The Netherlands, pp. 1383-1389.

17. Wormleaton, P.R., and Merrett, D.J. (1990). "An Improved Method of Calculation of Steady Uniform Flow in Prismatic Main Channel/Flood Plain Sections," *Journal of Hydraulic Research*, 28(2), pp. 157-174.

18. Myers, R.C., and Lyness, J.F. (1997). "Discharge Ratios in Smooth and Rough Compound Channels," *Journal of Hydraulic Engineering*, ASCE, 123(3), pp. 182-188.

19. Sturm, T.W., and Janjua, N.S. (1993). "Bridge Abutment Scour in a Floodplain," *Hydraulic Engineering '93, Proceedings of the Hydraulics Conference*, San Francisco, CA, ASCE, pp. 761-766.

20. Sturm, T.W., and Janjua, N.S. (1994). "Clear-Water Scour Around Abutments in Floodplains," *Journal of Hydraulic Engineering*, ASCE, 120(8), pp. 956-972.

21. Shearman, J.O. (1990). *User's Manual for WSPRO–A Computer Model for Water Surface Profile Computations*, Report No. FHWA-IP-89-027, FHWA, USDOT, Washington, DC.

22. Sadiq, Aftab (1994). *Clear-Water Scour Around Bridge Abutments in Compound Channels*, Ph.D. Dissertation, School of Civil and Environmental Engineering, Georgia Institute of Technology, Atlanta, GA.

23. Sturm, T.W., and Sadiq, Aftab (1996a). "Clear-Water Scour Around Bridge Abutments Under Backwater Conditions," *Transportation Research Record 1523*, TRB, National Research Council, Washington, DC, pp. 196-202.

24. Melville, B.W. (1997). "Pier and Abutment Scour: Integrated Approach," *Journal of Hydraulic Engineering*, ASCE, 123(2), pp. 125-136.

25. Sellin, R.H.J. (1964). "A Laboratory Investigation Into the Interaction Between the Flow in the Channel of a River and That Over Its Flood Plain," *La Houille Blanche*, 7, pp. 793-801.

26. Zheleznyakov, G.V. (1971). "Interaction of Channel and Floodplain Streams," *Proceedings of the 14th International Association for Hydraulic Research (IAHR) Conference*, Paris, France, IAHR, Delft, The Netherlands, pp. 145-148.

27. Tominaga, A., and Nezu, I. (1991). "Turbulent Structure in Compound Open-Channel Flows," *Journal of Hydraulic Engineering*, ASCE, 117(1), pp. 21-41.

28. Wright, R.R., and Carstens, M.R. (1970). "Linear Momentum Flux to Overbank Sections," *Journal of the Hydraulics Division*, ASCE, 96(9), pp. 1781-1793.

29. Yen, C.L., and Overton, D.E. (1973). "Shape Effects on Resistance in Floodplain Channels," *Journal of the Hydraulics Division*, ASCE, 99(1), pp. 219-238.

30. Wormleaton, P.R.; Allen, J.; and Hadjipanos, P. (1982). "Discharge Assessment in Compound Channel Flow," *Journal of the Hydraulics Division*, ASCE, 108(9), pp. 975-993.

31. Knight, D.W., and Demetriou, J.D. (1983). "Floodplain and Main Channel Flow Interaction," *Journal of Hydraulic Engineering*, ASCE, 109(8), pp. 1073-1092.

32. Wormleaton, P.R., and Hadjipanos, P. (1985). "Flow Distribution in Compound Channels," *Journal of Hydraulic Engineering*, ASCE, 111(2), pp. 357-361.

33. Radojkovic, M., and Djordjevic, S. (1985). "Computation of Discharge Distribution in Compound Channels," *Proceedings of the 21st Congress of IAHR*, Melbourne, Australia, IAHR, Delft, The Netherlands, 3, pp. 367-371.

34. Ackers, P. (1993). "Stage-Discharge Functions for Two-Stage Channels: The Impact of New Research," *Journal of the Institute of Water and Environmental Management*, 7(1), pp. 52-61.

35. Myers, W.R.C., and Brennan, E.K. (1990). "Flow Resistance in Compound Channels," *Journal of Hydraulic Research*, 28(2), pp. 141-155.

36. Blalock, M.E., and Sturm, T.W. (1981). "Minimum Specific Energy in Compound Open Channel," *Journal of the Hydraulics Division*, ASCE, 107(6), pp. 699-717.

37.   Chaudhry, M. Hanif, and Bhallamudi, S. Murty (1988). "Computation of Critical Depth in Symmetrical Compound Channels," *Journal of Hydraulic Research*, 26(4), pp. 377-396.

38.   Blalock, M.E., and Sturm, T.W. (1983). Closure to discussion of "Minimum Specific Energy in Compound Open Channel," *Journal of the Hydraulics Division*, ASCE, 109(3), pp. 483-487.

39.   Yuen, K.W.H., and Knight, D.W. (1990). "Critical Flow in a Two-Stage Channel," *Proceedings of the International Conference on River Flood Hydraulics*, Wallingford, U.K., edited by W.R. White, John Wiley & Sons, pp. 267-276.

40.   Sturm, T.W., and Sadiq, Aftab (1996b). "Water Surface Profiles in Compound Channel With Multiple Critical Depths," *Journal of Hydraulic Engineering*, ASCE, 122(12), pp. 703-709.

41.   Rastogi, A.K., and Rodi, W. (1978). "Predictions of Heat and Mass Transfer in Open Channels," *Journal of the Hydraulics Division*, ASCE, 104(HY3), pp. 397-420.

42.   Keller, R.J., and Rodi, W. (1988). "Prediction of Flow Characteristics in Main Channel/Floodplain Flows," *Journal of Hydraulic Research*, 26(4), pp. 425-441.

43.   Patankar, S.V., and Spalding, D.B. (1972). "A Calculation Procedure for Heat, Mass, and Momentum Transfer in 3D Parabolic Flows," *International Journal of Heat and Mass Transfer*, 15, pp. 1787-1806.

44.   Krishnappan, B.G., and Lau, Y.L. (1986). "Turbulence Modeling of Flood Plain Flows," *Journal of Hydraulic Engineering*, ASCE, 112(4), pp. 251-266.

45.   Prinos, P. (1990). "Turbulence Modeling of Main Channel-Floodplain Flows With an Algebraic Stress Model," *Proceedings of the International Conference on River Flood Hydraulics*, Wallingford, U.K., edited by W.R. White, John Wiley & Sons, pp. 173-186.

46.   Naot, D.; Nezu, I.; and Nakagawa, H. (1993). "Hydrodynamic Behavior of Compound Rectangular Open Channels," *Journal of Hydraulic Engineering*, ASCE, 119(3), pp. 390-408.

47.   Pezzinga, G. (1994). "Velocity Distribution in Compound Channel Flows by Numerical Modeling," *Journal of Hydraulic Engineering*, ASCE, 120(10), pp. 1176-1198.

48.   Chapman, R.S., and Kuo, C.Y. (1985). "Application of the Two-Equation K-ε Turbulence Model to a Two-Dimensional, Steady, Free Surface Flow Problem With Separation," *International Journal of Numerical Methods in Fluids*, 5, pp. 257-268.

49.   Puri, A.N., and Kuo, C.Y. (1985). "Numerical Modeling of Subcritical Open Channel Flow Using the K-ε Turbulence Model and the Penalty Function Finite Element Technique," *Applied Mathematical Modeling*, 9, pp. 82-88.

50.   Tingsanchali, T., and Maheswaran, S. (1990). "2-d Depth-Averaged Computation Near a Groyne," *Journal of Hydraulic Engineering*, ASCE, 116(1), pp. 71-86.

51.   Tingsanchali, T., and Rahman, K.R. (1992). "Comparison of Pseudo-Viscosity Model and K-ε Turbulence Model in 2-d Depth-Averaged Flow Computations," *Computer Techniques and Applications, Hydraulic Engineering Software IV*, edited by W.R. Blain and E. Cabrera, Elsevier Applied Science, 1, pp. 189-200.

52.   Khan, K.W., and Chaudhry, M.H. (1992). "Numerical Modeling of Flow Around Spur Dikes," *Computer Techniques and Applications, Hydraulic Engineering Software IV*, edited by W.R. Blain and E. Cabrera, Elsevier Applied Science, 1, pp. 223-235.

53.   Shen, H.W.; Chan, C.T.; Lai, J.S.; and Zhao, D. (1993). "Flow and Scour Near an Abutment," *Hydraulic Engineering '93, Proceedings of the 1993 Hydraulics Conference*, San Francisco, CA, ASCE, 1, pp. 743-748.

54.   Biglari, B. (1995). *Turbulence Modeling of Clear-Water Scour Around Bridge Abutment in Compound Open Channel*, Ph.D. Dissertation, School of Civil and Environmental Engineering, Georgia Institute of Technology, Atlanta, GA.

55.   Biglari, B., and Sturm, T.W. (1998). "Numerical Modeling of Flow Around Bridge Abutments in Compound Channel," *Journal of Hydraulic Engineering*, ASCE, 124(2), pp. 156-164.

56.   Ahmed, M. (1953). "Experiments on the Design and Behavior of Spur Dikes," *Proceedings of the International Hydraulics Convention*, ASCE and IAHR, University of Minnesota, Minneapolis, MN, pp. 145-159.

57.   Laursen, E.M., and Toch, A. (1953). "A Generalized Model Study of Scour Around Bridge Piers and Abutments," *Proceedings of the International Hydraulics Convention*, ASCE and IAHR, University of Minnesota, Minneapolis, MN, pp. 123-131.

58.   Garde, R.J.; Subramanya, K.; and Nambudripad, K.D. (1961). "Study of Scour Around Spur Dikes," *Journal of the Hydraulics Division*, ASCE, 87(HY6), pp. 23-37.

59.   Liu, H.K.; Chang, F.M.; and Skinner, M.M. (1961). *Effect of Bridge Construction on Scour and Backwater*, Report CER 60 HKL 22, Department of Civil Engineering, Colorado State University, Fort Collins, CO.

60.   Gill, M.A. (1972). "Erosion of Sand Beds Around Spur Dikes," *Journal of the Hydraulics Division*, ASCE, 98(HY9), pp. 1587-1602.

61. Laursen, E.M. (1960). "Scour at Bridge Crossings," *Journal of the Hydraulics Division*, ASCE, 86(HY2), pp. 39-54.

62. Tey, C.B. (1984). *Local Scour at Bridge Abutments*, Research Report No. 329, School of Engineering, University of Auckland, Auckland, New Zealand.

63. Dongol, D.M.S. (1994). *Local Scour at Bridge Abutments*, Research Report No. 544, School of Engineering, University of Auckland, Auckland, New Zealand.

64. Melville, B.W., and Ettema, R. (1993). "Bridge Abutment Scour in Compound Channels," *Hydraulic Engineering '93, Proceedings of the Hydraulics Conference*, San Francisco, CA, ASCE, pp. 767-772.

65. Young, G.K.; Palaviccini, M.; and Kilgore, R.T. (1993). "Scour Prediction Model at Bridge Abutments," *Hydraulic Engineering '93, Proceedings of the Hydraulics Conference*, San Francisco, CA, ASCE, pp. 755-760.

66. Young, G.K.; Dou, X.; Saffarinia, K.; and Jones, J.S. (1998). "Testing Abutment Scour Model," *Proceedings of Water Resources Engineering, '98*, Memphis, TN, ASCE, pp. 180-185.

67. Lim, S.Y. (1997). "Equilibrium Clear-Water Scour Around an Abutment," *Journal of Hydraulic Engineering*, ASCE, 123(3), pp. 237-243.

68. Chang, Fred F.M. (1996). "Abutment Scour Based on Flow Distribution Through a Bridge Opening," preprint of presentation at the North American Water and Environment Congress, Anaheim, CA, ASCE.

69. Palaviccini, M. (1993). *Scour Predictor Model at Bridge Abutments*, Ph.D. Dissertation, Civil Engineering Department, Catholic University of America, Washington, DC.

70. Chang, F., and Davis, S. (1998). "Maryland Procedure for Estimating Scour at Bridge Abutments, Part 2–Clear Water Scour," *Proceedings of Water Resources Engineering, '98*, ASCE, Memphis, TN, pp. 169-173.

71. Neill, C.R., editor (1973). *Guide to Bridge Hydraulics*, Roads and Transportation Association of Canada, University of Toronto Press, Toronto, Canada.

72. Rajaratnam, N., and Nwachukwu, B.W. (1983a). "Flow Near Groyne-Like Structures," *Journal of Hydraulic Engineering*, ASCE, 109(3), pp. 463-480.

73. Rajaratnam, N., and Nwachukwu, B.W. (1983b). "Erosion Near Groyne-Like Structures," *Journal of Hydraulic Research*, IAHR, 21(4), pp. 227-287.

74. Lim, S.Y., and Cheng, N.S. (1998). "Prediction of Live-Bed Scour at Bridge Abutments," *Journal of Hydraulic Engineering*, ASCE, 124(6), pp. 635-638.

75. Julien, P.J. (1995). *Erosion and Sedimentation*, Cambridge University Press, New York, NY.

76. Richardson, J.R. (1998). Discussion of "Equilibrium Clear-Water Scour Around an Abutment," by S.Y. Lim, *Journal of Hydraulic Engineering*, ASCE, 124(10), pp. 1071-1072.

77. Richardson, E.V.; Simons, D.B.; and Julien, P. (1990). *Highways in the River Environment*, Report No. FHWA-HI-90-016, FHWA, USDOT, Washington, DC.

78. Sturm, T.W. (1998). *Effect of Compound Channel Hydraulics on Bridge Abutment Scour*, Research Project 9411 Final Report, Office of Materials and Research, Georgia DOT.

79. Vanoni, Vito A., editor (1977). *Sedimentation Engineering*, ASCE, Reston, VA.

80. Carstens, M.R. (1966). "Similarity Laws for Localized Scour," *Journal of the Hydraulics Division*, ASCE, 92(3), pp. 13-36.

81. Raudkivi, A.J., and Ettema, R. (1983). "Clear-Water Scour at Cylindrical Piers," *Journal of Hydraulic Engineering*, ASCE, 109(3), pp. 338-350.

82. Yanmaz, A.M., and Altinbilek, H.D. (1991). "Study of Time-Dependent Local Scour Around Bridge Piers," *Journal of Hydraulic Engineering*, ASCE, 117(10), pp. 1247-1268.

83. Kothyari, U.C.; Garde, R.J.; and Rangu Raju, K.G. (1992). "Temporal Variation of Scour Around Circular Bridge Piers," *Journal of Hydraulic Engineering*, ASCE, 118(8), pp. 1091-1106.

84. Chiew, Y., and Melville, B.W. (1999). "Temporal Development of Local Scour at Bridge Piers," *Proceedings of the North American Water and Environment Congress* (on CD-ROM), Anaheim, CA, ASCE.

85. Pagan-Ortiz, J.E. (1991). *Stability of Rock Riprap for Protection at the Toe of Abutments Located at the Floodplain*, Report No. FHWA-RD-91-057, FHWA, USDOT, Washington, DC.

86. Sturm, T.W., and Chrisochoides, A. (1998). "One-Dimensional and Two-Dimensional Estimates of Abutment Scour Prediction Variables," *Transportation Research Record 1647*, TRB, National Research Council, Washington, DC, pp. 18-26.

87. Sturm, T.W., and Chrisochoides, A. (1998). "Abutment Scour in Compound Channels for Variable Setbacks," *Water Resources Engineering '98, Proceedings of the International Water Resources Engineering Conference*, ASCE, Memphis, TN, 1, pp. 174-179.

88. Froehlich, D.C. (1996). "Contraction Scour at Bridges: Analytic Model for Coarse-Bed Channels," *Proceedings of the North American Water and Environment Congress* (on CD-ROM), ASCE, Anaheim, CA.

89. Sheppard, D.M. (1999). "Conditions of Maximum Local Structure-Induced Sediment Scour," *Stream Stability and Scour at Highway Bridges, Stream Stability and Scour at Highway Bridges*, edited by E.V. Richardson and P.F. Lagasse, ASCE, pp. 347-364.

90. van Rijn, L. (1984). "Sediment Transport, Part I: Bed Load Transport," *Journal of Hydraulic Engineering*, ASCE, 110(10), pp. 1431-1456.

91. Brownlie, W.R. (1981). *Prediction of Flow Depth and Sediment Discharge in Open Channels*, Report No. KH-R-43A, W.M. Keck Laboratory, California Institute of Technology, Pasadena, CA.

92. Karim, M.F., and Kennedy, J.F. (1990). "Menu of Coupled Velocity and Sediment-Discharge Relations for Rivers," *Journal of Hydraulic Engineering*, ASCE, 116(8), pp. 978-996.

93. Mueller, D.S., and Hitchcock, H.A. (1998). *Proceedings of Water Resources Engineering*, '98, ASCE, Memphis, TN, pp. 210-215.

94. Jones, J. Sterling (1998). Personal communication, Turner-Fairbank Highway Research Center, FHWA, USDOT, McLean, VA, March.

95. Dou, Xibing (1998). Personal communication, GKY Associates, Springfield, VA, October.

96. *Bridge Scour Investigation for Bridge 5359 on Trunk Highway 12 Over Pomme de Terre River* (1995). Prepared by BRW, Inc., Minneapolis, MN, for MinnDOT.